AF614682

Open-Source Horizons - Challenges and Opportunities for Collaboration and Innovation

Edited by Laura M. Castro

Published in London, United Kingdom

Open-Source Horizons - Challenges and Opportunities for Collaboration and Innovation
http://dx.doi.org/10.5772/intechopen.111234
Edited by Laura M. Castro

Contributors
Adilson Luiz Pinto, Alexandre Ribas Semeler, Dharmender Salian, Edson Mário Gavron, Fabio Lorensi do Canto, Laura M. Castro, Martin Hristev, Mayukh Sarkar, Pawel Gasiorowski, Reza Baghaeishiva, Sabin Nakarmi, Sorin Radu, Sruti Biswas, Tarun Bali, Vassil Vassilev, Viktor Sowinski-Mydlarz

First published in London, United Kingdom, 2024 by IntechOpen
IntechOpen is the global imprint of INTECHOPEN LIMITED, registered in England and Wales, registration number: 11086078, 167-169 Great Portland Street, London, W1W 5PF, United Kingdom

British Library Cataloguing-in-Publication Data
A catalogue record for this book is available from the British Library

Additional hard and PDF copies can be obtained from orders@intechopen.com

Open-Source Horizons - Challenges and Opportunities for Collaboration and Innovation
Edited by Laura M. Castro
p. cm.
Print ISBN 978-0-85466-113-8
Online ISBN 978-0-85466-112-1
eBook (PDF) ISBN 978-0-85466-114-5

For EU product safety concerns:
IN TECH d.o.o., Prolaz Marije Krucifikse Kozulić 3, 51000 Rijeka, Croatia,
info@intechopen.com or visit our website at intechopen.com.

Meet the editor

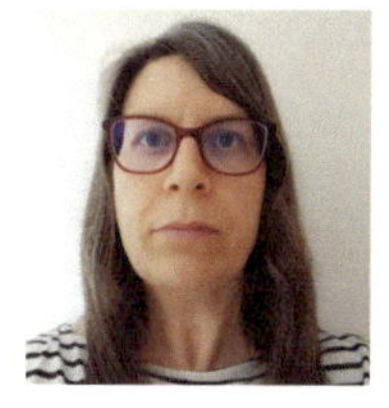

Dr. Laura M. Castro is a professor at the University of A Coruña, Spain, where she has been lecturing for more than 15 years. Apart from being a coordinator of several undergraduate courses on software architecture and software testing, she is currently the president of the Institutional Chair for Open Science Promotion through Software (CICAS). Her most recent research interests focus on the automatic validation of distributed systems. She has supervised three Ph.D. theses and acted as principal investigator for several European projects. She is also actively involved in dissemination activities, particularly the visibility of women in STEM, as a member of the Association for Computing Machinery's Council on Women in Computing (ACM-W) Europe.

Contents

Preface

Welcome to *Open-Source Horizons – Challenges and Opportunities for Collaboration and Innovation*. In this edited volume, we aim to present the reader with a panoramic view of different aspects of the dynamic landscape of open-source development, exploring the intersections of technology, community, and creativity.

The open-source movement has transcended mere software development; it has become a philosophy, a way of thinking that embraces transparency, inclusivity, and collective progress. In the chapters ahead, expect to encounter diverse perspectives - from seasoned practitioners to curious newcomers, from experienced technical researchers to knowledgeable humanistic experts - each contributing to the ever-expanding horizon of open collaboration.

Our contributors address different challenges in which open-source is a cornerstone, and those challenges are as relevant today as ever. The reader will be confronted with questions that range from sustainability (have the open data initiatives, which were expected to bring transparency and citizen empowerment to our democratic societies, lived up to said expectations?) to sovereignty (how can we use open source to retain sovereignty over our data and processes?), and featuring of course the intersection with the AI revolution we seem to be currently immersed in.

Our efforts in creating this volume stem from the belief that within these challenges lie remarkable opportunities. The open-source ecosystem fosters innovation, empowers individuals to shape technology, and democratizes access to knowledge and tools. As you turn these pages, I hope you find not only interesting insights but also inspiration. Whether you're a developer, a designer, a researcher, or just a curious person, this book is for you. May this volume inspire you to explore new horizons, collaborate boldly, and contribute to the ever-evolving tapestry of open-source horizons.

Dr. Laura M. Castro
Department of Computer Science and Information Technologies,
Chair of Cátedra por el Impulso a la Ciencia Abierta a través del Software (CICAS),
University of A Coruña,
A Coruña, Spain

Chapter 1

Introductory Chapter: Unveiling Present and Future Horizons of Open Source

Laura M. Castro

1. Introduction

The roots of the Open Source movement can be traced back to the early days of computing [1], with the advent of freely shared software among the pioneering community of programmers. However, it was not until the 1980s that the term "open source" was coined and popularized: a group of developers led by Linus Torvalds created the Linux operating system, which was released as free and open source software. This project paved the way for the Open Source movement by demonstrating the power and potential of collaborative software development [2]. The Open Source movement continued to gain traction in the enterprise world in the early 2000s, with companies like IBM, Red Hat, and Oracle offering open source solutions for business applications, databases, and operating systems [2]. Today, Open Source lives a new revolution, in which the cornerstone role of software in societal advance is being embraced as well in the scientific arena, and has paired the term with others like Open Data, Open Access, and Open Science.

1.1 The tapestry of the open movements

The practical benefits of collaborative work and exploration are without discussion [3]. Indeed, they extend beyond software, giving rise to parallel movements in various domains. First, the Open Data [4] movement emerged, claiming for the unrestricted availability and sharing of data. Recognizing the transformative power of data in driving innovation, transparency, and accountability, this movement has influenced governments, research institutions, and businesses to adopt open data practices. It also addresses the empowerment of citizens toward gaining control over their own data.

In tandem, Open Access [5] appeared, another consequential offshoot, which focuses on removing barriers to scholarly publications. Scholars and researchers advocate for unrestricted access to academic literature, ensuring that knowledge is not confined by paywalls. Open Access has redefined the dissemination of research findings, fostering a more egalitarian access to information.

This path has led to Open Science [6], a philosophy advocating for the transparent and collaborative sharing of scientific knowledge, thus aligning closely with the ethos of the Open Source movement. The goal is to make scientific research more accessible, reproducible, and inclusive. Collaborative platforms, open-access journals, and initiatives promoting data sharing have become integral components of the Open Science movement, reshaping the landscape of academic research. Institutions like the European Union have reformulated their funding programs to adopt strong Open Science commitments [7].

These interconnected movements, often referred to collectively as "the Open Movements", form a collaborative tapestry that extends beyond the realm of technology. As technology, science, data, and knowledge intertwine, the Open Movements collectively shape a paradigm that emphasizes collaboration, inclusivity, and the democratization of information. The success of one movement often paves the way for advancements in others, creating a synergistic ecosystem that has become a hallmark of progressive and inclusive practices in the XXI century. The evolution of these movements reflects a commitment to openness and underscores the transformative power of collaborative efforts in shaping the future of our interconnected world.

2. Purpose and goals

In editing this book, a clear mission was set: to provide a state-of-the-art look at the present and future of Open Source. This is not a technical manual; it is a mosaic of perspectives that examines Open Source from various vantage points. We aim to illuminate its critical role in Open Science, its strategic significance for small businesses, its democratic potential within public administrations, and the nuanced implications revealed through a societal-focused analysis. We also pursued the exploration of its imminent coexistence with the emergent large AI-powered systems.

Our ultimate purpose is twofold: to provide a sneak peek of the fringe of Open Source, and to present a forward-looking panorama of its far-reaching implications from a comprehensive perspective, to foster critical thinking about the societal impact of the Open Movements. By doing so, we invite readers to adopt a holistic viewpoint, understanding the intricate threads that weave the fabric of Open Source into the broader societal canvas.

3. Conclusion

At its core, this book goes far beyond lines of code and celebrates the spirit of collaboration that defines the Open Source community. It serves as a testament to the collective ingenuity that has propelled Open Source to the forefront of technological progress. Through case studies and real-world examples, we showcase the accomplishments and potential of a community driven by a shared vision for a more open, accessible, and inclusive digital future.

Whether the reader is a seasoned technologist, a business strategist, a public servant, or an enthusiast curious about the digital realm, this volume hopes to be an inspiring resource. The Open Source phenomenon is not confined to the realm of technology—it transcends its technological roots and has become a cultural, societal, and transformative force. By offering diverse perspectives, we hope to inspire further thinkers, builders, and advocates who will shape the trajectory of open source in the years to come.

With "Open-Source Horizons - Challenges and Opportunities for Collaboration and Innovation", we extend an invitation to engage in an exploration, where the meticulous examination of collaborative innovation converges with intellectual rigor to unravel the complexities and prospectives of Open Source.

Author details

Laura M. Castro
University of A Coruña, A Coruña, Spain

*Address all correspondence to: lcastro@udc.es

References

[1] Brasseur VM. Forge your Future with Open Source. Raleigh, North Carolina (US): Pragmatic Programmers; 2018. ISBN 978-1-68050-301-2

[2] Michels M. A Brief History of Open Source. 2021. Available from: https://maximilianmichels.com/2021/history-of-open-source/ [Accessed: November 13, 2023]

[3] Kilamo T, Hammouda I, Chatti MA. Teaching collaborative software development: A case study. In: 34th International Conference on Software Engineering (ICSE). Zurich, Switzerland: IEEE; 2012. pp. 1165-1174. DOI: 10.5555/2337223.2337376

[4] Open Data Handbook. Available from: https://opendatahandbook.org/guide/en/what-is-open-data/ [Accessed: November 13, 2023]

[5] Dutch National Open Access Website: What is open access? Available from: https://www.openaccess.nl/en/what-is-open-access [Accessed: November 13, 2023]

[6] UNESCO. Open Science: Making Science more Accessible, Inclusive and Equitable for the Benefit of All. Paris, France: UNESCO; 2019. Available from: https://www.unesco.org/en/open-science [Accessed: November 13, 2023]

[7] European Commission. The EU's Open Science Policy. Brussels, Belgium: European Commission; 2019. Available from: https://research-and-innovation.ec.europa.eu/strategy/strategy-2020-2024/our-digital-future/open-science_en [Accessed: November 11, 2023]

Chapter 2

Usability of Open Data

Dharmender Salian

Abstract

Open data (OD) is the term used to describe the concept that data is available freely for people, entrepreneurs, and researchers for analysis and research. Globally, Governments have taken initiatives to publish public data. Researchers and entrepreneurs wanting to do data analysis need to be trained in data management as the quality and accessibility of open data datasets make the activity a bit challenging. OD initiatives require considerable resources that include financial, technical, and human resources. Using concepts of structuredness of data, a dataset usability measurement is created. Utilizing a randomly chosen set of datasets from a well-known open data portal, an instrument is developed, validated, and applied. The chapter ends with explaining future research directions and giving recommendations for distributors of open data datasets.

Keywords: usability, quality, reuse, data, citizen

1. Introduction

On 7 December 2007, a meeting was held with 30 individuals in Sebastopol, California, to review how to define "open public data" [1]. This was the initial effort to define "open Public data". They wrote eight principles of open government data.

- Complete: it has relevant details
- Primary: data is gathered at the source
- Timely: data is made available soon after collecting
- Accessible: data is convenient for a wide range of users and purposes.
- Machine processable: automatic processing is possible.
- Non-discriminatory: any person can use it
- Non-proprietary: not protected by any enterprise
- License-free: data is not regulated.

A year later, the memoranda on open data were signed by US President Obama. Open data is data that anybody can use, access, and share. Open data turns into usable while made accessible in a general machine-readable format. People should be allowed to use the data as they want, including sharing or transforming. There are questions about open data vs. open government data vs. public data. Open government data is open data when government influence is minimal regarding the re-use of data [2]. Open data is available to the public, entrepreneurs, and researchers so that they can create new services and products and have successful businesses [3]. There are benefits of open data which can vary from commercial value through innovation to increased government effectiveness. Entrepreneurs using open data help in the open data movement and help in exploiting the full potential of open data. The public should be encouraged to use open data and turn ideas into successful businesses. New insights and knowledge can be obtained from open data by making it easily accessible and transparent.

There is a need to involve researchers, students, and citizens to promote open data and boost open data development. Open data is available in portals and interoperability is most important. Interoperability denotes the ability of various systems and groups to work collectively (inter-operate). In this situation, it's far the capacity to interoperate—or intermix—different datasets. Interoperability means the capability to access data from two or more sources and integrate that data for further usage. Interoperability is essential because it permits different components to work together. In some portals, data is of low quality, has sensitive information, and has issues with interoperability [4].

Governments globally are taking initiatives like Data.gov and Data.gov.uk. These initiatives are helping the goals of the open data movement. The open government suggests the government is collaborative and efficient [5]. Linked open data is also open data that is linked data or structured data, which is interlinked with other data, so it turns beneficial via semantic queries. Semantic queries assist data procurement in a programmatic fashion. Public or private organizations can often control access or re-use of data through the license, and this acts as a barrier to open data movement. Organizations have a license to ascertain the status of the data set and otherwise, this may limit the use of the open data. Open data is about being accessible, available, and reusable. For an organization, data availability refers to how often data is available to be used. Entrepreneurs are mostly developing products and services. This will require reusable data so it can be transformed as per the requirements. Data accessibility is the degree to which government data is supplied in open and reusable formats, along with associated metadata. Though a lot of open government data is available on portals, awareness of its existence and usefulness is limited to citizens hence need to raise awareness [6].

Data transparency is the attribute of data being used with integrity, and lawfully, for legitimate purposes. Increased accountability and decreased corruption are the advantages of increased transparency. Detection and reduced corruption are found if scrutiny and monitoring are given their due prominence. Organization officials will be averse to doing anything wrong if they know that they are being watched. Inspection at every stage should be mandatory in organizations. Officials should encourage transparency from top to bottom. Increased transparency needs clarity and discussion with other officials so all know where the boundaries are, and anything beyond will be punished. Transparency leads to increased trust and better employee engagement. Suspicion of illegal activity and bad practices can lead to barriers built up against an organization and its products which may lead to a boycott. Trust is essential for an organization. Trust can be from employees, customers, vendors, and other stakeholders. Good relationships and long-term bonds are possible with trust and good dealings. In today's world news moves fast whether on social media or other sources and it can make or break an organization's reputation in the market. Customers value dealing with organizations that have a clean chit and are ethically business minded. Transparency in internal workings and in business dealings with customers and vendors will lead to better relations with all stakeholders.

Open data becomes useful when a machine can utilize it and human can understand it. Thus, open data quality is important. Open data quality needs to have five characteristics, Accuracy, Completeness, Reliability, Relevance, and Timeliness [7]. Accuracy indicates that the information is correct in every respect. Completeness shows how all-inclusive is the information. Reliability shows whether the information is disputed in any other dependable source. Relevance gives the necessity for it. Finally, Timeliness tells whether the information is contemporary. Data quality indicates whether the data can be utilized in requisite circumstances. When data quality is meeting these requirements then it is valuable and can be utilized in the required context.

Open data reuse is dependent on data being in the required format and in procedures that allow data to be used by any person. If open data is available in the required format so that any entrepreneur must make a minimum effort before reusing it, then it is going to be valued more. If extra input work is required to bring it to the needed form, then it will set up a barrier to open data usage. There is a need that open data portals should allow the reuse of open data, ensure the efficiency of data transmission, and enable professional initiatives based on data reuse [8].

Governments are becoming responsive to this data and are having innovative programs to cater to open data. Globally new portals have been started to cater to open data and initiatives like Data.gov and Data.gov.uk. Open government data is a form of open data. This is created by government institutions. Open data may include non-textual material like maps and even medical data. The commercial value of data has acted as a barrier to open data initiatives and licenses, charges are to be paid to reuse data. This definitely hampers the progress of the open data movement and people suggest license-free will be for the good of all concerned. Open data can improve government functioning due to transparency and less corruption. It can also bring in new tools to solve societal or real-world problems. People say that it was public money used to generate this data and restrictions on re-use are not required.

Governments have open data portals where public institutions' datasets are available in different formats. There is a need for ensuring data quality. Different Organization units roll out this open data and there is a lack of quality standards even within different units of one organization. Effective use of open data datasets will be hampered due to poor-quality data. Different standards of open data will slow down open data initiatives as consumers will be forced to waste time and resources to improve this data so it may be re-used for actual requirements. Quality in open data is measured by characteristics like completeness, accuracy, timeliness, relevance, and reliability. Open data quality is important as users will not waste their time and resources to improve it. Ensuring quality open data at the source or portal itself can ease problems faced by end users. Quality planning, assurance, and control need to get their due importance. Information when provided in a standardized format is known as structured data. There are three types of data for analysis. Structured, semi-structured, and unstructured. Semi-structured is easier to handle than unstructured data and is not formatted in conventional ways. CSV and XML are semi-structured documents as CSV files can be easily imported into SQL for further analysis.

File formats can be of the type, JSON, XML, RDF, Comma separated files, test documents, Plain text, HTML, etc. Choosing the correct format is essential so there is the highest usability of data. Portal managers should have open data in a standardized, machine-readable format that can be readily used by customers.

This paper is in different sections. Section 2 discusses the literature survey that is about open data datasets quality. Section 4 relates to the sampling of the datasets. Section 5 details the development of a tool to measure and evaluate the usability of open data. Section 7 deals with future possibilities for open data for researchers, entrepreneurs, and other stakeholders.

2. Literature survey

Data-driven global economy reflects the technological breakthroughs in the past few decades. Data has become the new oil. New tools are flooding the market to exploit the data. Collection and use of data are challenging and are becoming complex with technological advances all around. Global internet traffic has increased multifold and data transmission is huge. It is going to evolve over the coming decades. Entrepreneurs have seen this growth and know the potential this data economy has. Digital technologies hold enormous potential, and this is just the beginning. Creators of data are now looking at data and the value associated with it. Policies and strategies are needed to get value from data-driven opportunities.

Data processing and cleanup are required to obtain the hidden value. Technology solutions have opened new opportunities which if planned well can lead to value for organizations, society, and the nation. The quality of data is also relying on context and user [9]. Free Government datasets are available on the web on different platforms in a non-proprietary format [10]. For data sources, a process of sources assessment, determination of quality grades, and then the final selection needs to be specified [11].

Users want good-quality data when they are using it. The structure is important as records must have keys that are non-empty, easy to analyze, and trustable and formats must be machine-readable. Missing values, inconsistencies within a column, superfluous information, duplicates, and bad formatting can lead to losses as extensive data cleaning operations will be required. Hence open data quality needs to be measured and made as clean as possible, so it is ready for reuse requirements by users. These factors need to be eliminated to remove damaging usability possibilities in the future. High-quality data can be a thin dividing line between success and failure. To achieve these goals, structural usability assessment is required for open data datasets. Organizations need a metric as there are innovative technologies and data processing techniques in the market, but these need good quality data to provide useful results after re-use. There are different file formats available like CSV, XLSX, ZIP, TXT, JSON, JPEG, PNG, ZIP, HTML, etc. and there is a need to be able to change from one format to another as per user requirement. Most software programs for data analysis can use different formats like XML-based files (XML), text files, comma-separated values (CSV), spreadsheets, rich text formats (RTF), and database tables.

Open data needs a metric that will be a measuring system to evaluate the quality of open data. This will establish how useful and relevant the dataset is. This will separate high-quality data from poor-quality data based on this metric. There are two established methods, which are the Global Open Data Index and Open Data Barometer. These are discussed below separately due to their usefulness and prominence. Also, there is another index, Open Data Inventory which is the first index to evaluate both the coverage and openness of national statistical systems [12]. The focus of this index is to discover gaps, promote open data guidelines, enhance access, and inspire communication among national statistical offices (NSOs) and data users. The Open Data Inventory pivots on macrodata. Survey responses and administrative records are the eventual source for most microdata, a unit record level for macrodata. The Open Data Inventory measures how thorough a country's statistical offerings are and whether its records meet global standards of openness.

3. Available indexes for open data

Open data indices indicate how open a data portal is and encourage open data policies. We will discuss two open data indices. Open data barometer and global open data index. Both these evaluate the openness of an open dataset based on certain questions which vary from whether the data exists, is publicly available, whether payment is required, bulk availability, Is data machine-readable, and timeliness of data.

3.1 Global open data index

It shows the state of open government data publication. It is an independent assessment and tracks governments' progress on open data release. The scope is narrow and confined to practical openness. The effort is not made about use or impact. Data quality, which is a significant barrier to re-use, is not covered. There are a set of questions in the survey and scoring is done based on feedback of the replies received. After results are published then feedback is obtained from public officials to improve assessment. It is a challenging effort and is focused on governments so they can improve their scores. Open Data Index uncovers a (simple) API for programmatic access to data. Currently, the API is accessible in both JSON and CSV formats. The Open Data Index has a small set of tools and patterns for enforcing visualizations. Data is made ready by the Python script. This pulls data from the Open Data Index Survey (Census) and develops it in diverse ways after which writes it to the data directory.

3.2 Open data barometer

It evaluates the occurrence and global impact of open data initiatives. It looks at governments' usage of open data for social impact and accountability. The survey is done with questions on open data, policy, and implementation. It is a key benchmark regarding progress on openness and change policies in more than a dozen countries. It inspects world trends, and supplies country-wise data focused on open data impact and readiness.

But as shown above, limited data is gathered on the structural usability of open data datasets. Criteria used in these indices are based on functional dependencies and data summaries. This document will focus on the structural accuracy of the open data datasets.

4. Sampling the datasets

Here the effort is made to develop a tool and metric to be useful for the assessment of open data datasets. There are numerous government portals for open data (see **Table 1**). Most portals have a few thousand datasets as these are from government agencies and organizations. The US government's data.g ov portal has a few lakhs of datasets that are from the Federal government, local government, and nonfederal open data resources. The intention is to provide access to open government data, boost innovation and have a transparent government. Data.gov portal datasets were downloaded to collect samples. These were random samples of CSV data files. These datasets were analyzed further to build a score that can help in the quantitative evaluation of these datasets. A focused effort is made to find structural shortcomings in these files to help in the development of a tool.

No.	Country	Website
1	Argentina	https://datos.gob.ar/
2	Australia	https://data.gov.au/
3	Austria	https://www.data.gv.at/
4	Bangladesh	http://data.gov.bd/
5	Canada	https://open.canada.ca/en/open-data/
6	Denmark	https://www.opendata.dk/
7	Finland	https://www.avoindata.fi/en
8	France	https://www.data.gouv.fr/fr/
9	Germany	https://www.govdata.de/
10	India	https://data.gov.in/
11	Japan	https://www.data.go.jp/
12	Russia	https://data.gov.ru/
13	Saudi Arabia	https://data.gov.sa/en/
14	Spain	https://datos.gob.es/en
15	Sweden	https://www.dataportal.se/en/
16	Switzerland	https://opendata.swiss/
17	Taiwan	https://data.gov.tw/
18	Tanzania	http://opendata.go.tz/
19	Thailand	https://data.go.th/
20	Ukraine	https://data.gov.ua/
21	United Kingdom	https://data.gov.uk/
22	Uruguay	https://catalogodatos.gub.uy/

Table 1.
Open data portals of different countries.

5. Evaluating the usability of the datasets

In general, it is seen that data is in two forms, rectangular and non-rectangular. In rectangular, data is shaped as a rectangle with every data value analogous to some rows and columns. The sample files selected had headers and rectangular shapes. But a few files were in non-rectangular form also. Non-rectangular data are not systematically arranged. There can be inconsistent datatype in the same column or sometimes missing. Thus, the similarity of data structure is followed by a different one. Pivot data also is found as "[Commodity]", "[Commodity]. [All Commodities]. [Foods]. [Fried Goods]". This is useful data but becomes complex for analysts. Software tools like Power BI and Excel can be used to unpivot data. Details shown in **Table 2** discuss the dataset with a star schema. Data shown in **Table 3** discusses duplicates and inconsistent columns.

Based on the discussion above and the analysis, the dataset can be evaluated using these questions (see **Table 4**).

The weights are given based on importance. Also, among datasets, it was seen that variations occur due to distinct reasons like the dataset did not have a 'rectangular' shape, because it had a report header and many blank rows, and also the headers were not variable names but were instead variable values (see **Table 5**).

If a dataset has "Yes" for Question 1, it will get 0.20 points for Question 1. If it has "No" for Question 5, it will get −0.09 for Question 5. The sum of these points will be given to the utility index. Thus, for each dataset usability can be calculated. Weight can vary from the lowest 0–25% highest based on the importance each user assigns it as per his requirement. The summation of all weights is 100%. The dataset can have the highest score of 1. Samples taken from the US government's data.gov portal were found to have these criteria. 94% of the sample datasets downloaded were rectangular whereas 90% of the datasets had inconsistent column values (see **Table 6**).

Customer_id	Customer name	Expenditure	Stateid_fk	Cityid_fk	Countyid_fk	Subcountyid_fk
101	Davis	1021.00	1	2	5	7
102	Fred	1239.00	5	5	2	3
103	Smith	1892.00	3	5	8	9
104	Jones	1972.00	8	7	9	1

Table 2.
Dataset with star schema.

Inmate #	Inmate name	Sex	DOB	Release type	Release date
4319087721	John Dylan	M	08/21/1991	Misdemeanor	02/22/2020
2319087634	John Dylan	M	08/03/1990	Misdemeanor	07/15/2021
4317793234	Barbara Holster	F	01/05/1999	Supervised Release Program	03/17/2020
7819087634	Fred Wilson	Multiple	01/03/1980	Misdemeanor	05/19/2022
9311187634	Kiran Naveen	M	08/01/1981	Supervised Release Program	09/11/2020
5312387634	Jeevan Cook	M	08/02/1985	Supervised Release Program	02/15/2022

Table 3.
Dataset with an identifier column, duplicates, non-standard format, and inconsistent column.

No.	Questions	Yes	No	Weight
1	Is the dataset a rectangular table?	+	—	W1
2	Column values are consistent	+	—	W2
3	Follows star schema	+	—	W3
4	Identifier is present	+	—	W4
5	Grouping of rows	—	+	W5
6	Grouping of columns	—	+	W6
7	Existence of multiple dependencies among the columns	Zero	+	W7
8	Non-standard format	—	+	W8

Table 4.
Questions.

No.	Questions	Yes	No	Weight (%)
1	Is the dataset a rectangular table?	+	—	20
2	Column values are consistent	+	—	16
3	Follows star schema	+	—	12
4	Identifier is present	+	—	12
5	Grouping of rows	—	+	9
6	Grouping of columns	—	+	9
7	Existence of multiple dependencies among the columns	0	+	11
8	Non-standard format	—	+	11

Table 5.
Weights allocation.

No.	Questions	Yes (%)
1	Is the dataset a rectangular table?	94
2	Column values are consistent	90
3	Follows star schema	4
4	Identifier is present	80
5	Grouping of rows	15
6	Grouping of columns	11
7	Existence of multiple dependencies among the columns	32
8	Non-standard format	20

Table 6.
Datasets properties.

Using the utility index, scores of the sample datasets were calculated. All the sample scores for different sample datasets were aggregated and analyzed for further details (see **Table 7**).

Mean	0.700
Standard error	0.040
Median	0.800
Mode	0.900
Skewness	−1.9

Most datasets have high scores (Median = 0.800, Mean = 0.700). Also, there were a small number of low-scoring datasets (Skewness = −1.9).

Table 7.
Summary statistics of usability scores.

6. Limitations of this study

The research examined only datasets from a few portals like data.gov. In the future, to have more diverse datasets then can look at samples from portals of other countries. DataPortals.org shows that there are 598 portals on open data worldwide. A more diverse sample will show the problems with datasets globally. A list can be made country-wise which will detail the data quality of each of the portals worldwide. The US and many countries view the setting up of portals as being beneficial to society, but some others may view it as revenue generation, and still, others may do it for name's sake as will consider it too prohibitive for their country.

Details that can be gathered about these portals are whether they share good-quality data. Datasets need to be examined for compliance, consistency, completeness, and correctness. Portal provides datasets in which formats like CSV, JSON, and more. Is there an effort to repair datasets after receiving them from different organizations and before sharing them with the public? How many formats are totally provided so as to give a wider choice and should be humanly readable and machine-readable? Formats like CSV, JSON, and XML are more popular with the public and are easily shareable. The right structure and format are important to maximize the usability of data, ease of access for customers, and hidden cost of repairs of datasets for reuse.

Portals also need to look at the timeliness of datasets available on their site. Users need to make informed decisions or reuse these datasets for their services hence timeliness is important. The solution could be open data datasets that can be connected to the master database [13]. Thus, minimal update issues. Also, users will like that the portals automate the process of datasets being collected, processed, and stored as these if performed manually will be too laborious and losers will be users due to any slack in handling.

Design issues exist at many data portals. Datasets are provided as given by organizations and not as required by users. The user may be searching for a place, but the provider may not be giving a dataset in that format and hence irrelevant to the user. Data USA is a site that has looked at these design issues and improved its site. The effort has been to bring it closer to users and make data understandable to humans. It provides visualizations to help users and is principally for search engines. Through these efforts of data visualization and transforming data, users are happy, and it shows in millions of user numbers visiting the site [14]. Portal should help in understanding the structure, methodology, and arrangement of the datasets thus making life easier for users.

Based on the above details, a future list can be made of global open data portals and ranked accordingly. This could help users as they can know beforehand the portal data quality. Also, this could act as an incentive for portals to improve their data quality. Open data has gained popularity, but portals need to make it straightforward and easy for users. This global ranking of portals should be based on the structuredness of datasets, user-friendliness of the portal, formats of datasets available, and timeliness. These are some of the core features of the portals which will be beneficial to users as well as lead to a huge number of users on their site.

6.1 Future research directions

There is a need for automated tools for classifying and assessing the datasets available at the portals. Users usually take time to find and query the required datasets. Portals need to be looked at from factors such as speed, effectiveness, and satisfaction. Speed will be how quickly users can analyze the datasets. Effectiveness is whether users achieved the goals and satisfaction is about datasets fulfilling the requirements of users. Obtaining quality data is difficult and organizations are presently using tools like Talend Open Profiler, Apache Griffin, and power matchmaker. I have shown these names so it should be known that there are cleaning tools available in the market that can help users. These can be used for data validation, processing, and assessment. There is still a need for automated tools for open data so that users can spend less time searching for datasets and spend less on the hidden costs of repairing datasets.

7. Conclusion

This study is directed toward the perspective on the usability of open data. Citizens, entrepreneurs, and stakeholders are at the heart of this open data initiative. Designing an open data output is not enough but it needs to satisfy the needs of its end users. These include the structuredness of the open data dataset. Dataset use for decision-making, and reuse will be more popular if there is support for different formats, interoperability, and minimum costs in cleansing efforts to end users. Portals, public sector agencies, and other open data dataset producers need to ensure that values under columns have the same data type and minimal missing values. Users are wanting to reuse data and hence portals should look at improvements in a form so to perform visualizations and analytics is easier.

The effort in this chapter has been to shed more light on the structuredness of open data datasets and how measuring it will give the users a choice of choosing the dataset of any portal based on the requirement. The utility index is a good starting point to get details of the structuredness of open data datasets.

Appendix: definitions of terms

CSV (comma separated values) file

File is a text file that has a specific format that allows information to be saved in a table-based format.

Data

Data includes lists, tables, graphs, charts, and images. Data may be structured or unstructured and organized.

Data cleaning or scrubbing

Data cleansing, additionally known as data cleaning or data scrubbing, is the action of fixing incorrect, incomplete, duplicate or otherwise erroneous data in a data set.

Data portal

A portal is a web-based platform that collects information from distinct sources into a single user interface and presents users with the most applicable information for their context.

Database

Can be a software system for processing and managing data.

Dataset

A dataset is any organized collection of data.

File format

The file format refers to the internal arrangement (format) of the file, not how it is displayed to users. For example, CSV and XLS files are structured very differently, but may look similar or identical when opened in a spreadsheet program. The format corresponds to the last part of the file name or extension.

Machine readable

Able to be understood and used by a computer. To be machine readable, data must be structured in an organized way. CSV, JSON, and XML among others, are formats that contain structured data that a computer can automatically read and process.

Metadata

Metadata is information about a dataset that makes the data easier to find or identify.

Open data

Data is open if it can be freely accessed, used, modified and shared by anyone for any purpose.

PDF (portable document format)

PDF is a multi-platform file format.

Structured data

Structured data refers to information with a high degree of standardization, clearly defined and making the data readily searchable by search engines.

Unstructured data

Unstructured data (or unstructured information) is information that is usually stored in its native format. Either does not have a pre-defined data model or is not organized in a pre-defined manner, such as a flat file.

XML

Extensible Markup Language is a flexible file format designed to store, transport and share data over the Internet.

Author details

Dharmender Salian
University of Cumberlands, New York, USA

*Address all correspondence to: dsalian0302@ucumberlands.edu

References

[1] 10 Years of Open Data. Opendatasoft. 2017. Available from: https://www.opendatasoft.com/en/blog/open-data-anniversary-ten-years-after-the-sebastopol-meeting/

[2] Dickinson A. Whats the difference between open data and open government data. Medium. 2016. Available from: https://medium.com/@digidickinson/whats-the-difference-between-open-data-and-open-government-data-8a28eb525d2a

[3] Alzamil ZS, Vasarhelyi MA. A new model for effective and efficient open government data. International Journal of Disclosure and Governance. 2019; **16**(4):174-187. DOI: 10.1057/s41310-019-00066-w

[4] Bargh MS, Choenni S, Meijer R. ICEGOV '15-16: Proceedings of the 9th International Conference on Theory and Practice of Electronic Governance. New York, United States: Association for Computing Machinery; 2016. pp. 199-206. DOI: 10.1145/2910019.2910037

[5] Corrêa AS, Paula E Cd, Corrêa P, Pizzigatti L, Silva FS Cd. Transparency and open government data. Transforming Government: People, Process and Policy. 2017;**11**(1):58-78. DOI: 10.1108/TG-12-2015-0052

[6] Chokki AP, Simonofski A, Frénay B, Vanderose B. Open government data awareness: Eliciting citizens' requirements for application design. Transforming Government: People, Process and Policy. 2022;**16**(4): 377-390. DOI: 10.1108/TG-04-2022-0057

[7] Sarfin RL. 5 Characteristics of Data Quality. Precisely; 2022. Available from: https://www.precisely.com/blog/data-quality/5-characteristics-of-data-quality#:~:text=There%20are%20five%20traits%20that,read%20on%20to%20learn%20more

[8] Abella A, Ortiz-de-Urbina-Criado M, De-Pablos-Heredero C. Criteria for the identification of ineffective open data portals: Pretender open data portals. El Profesional De La Información. 2022;**31**(10). DOI: 10.3145/epi.2022.ene.11

[9] Martin EG, Law J, Ran W, Helbig N, Birkhead GS. Evaluating the quality and usability of open data for public health research: A systematic review of data offerings on 3 open data platforms. Journal of Public Health Management and Practice. 2017;**23**(4):e5-e13. DOI: 10.1097/PHH.0000000000000388

[10] D'Agostino M, Samuel N, Sarol M, de Cosio F, Marti M, Luo T, et al. Open data and public health. Revista Panamericana de Salud Pública. 2018;**42**. DOI: 10.26633/RPSP.2018.66

[11] Stróżyna M, Eiden G, Abramowicz W, Filipiak D, Małyszko J, Węcel K. A framework for the quality-based selection and retrieval of open data—A use case from the maritime domain. Electronic Markets. 2018;**28**(2):219-233

[12] Assessing the Coverage and Openness of Official Statistics. Open data Inventory. Open Data Watch. n.d. Available from: http://opendatawatch.com/monitoring-reporting/open-data-inventory/

[13] Schrack A. Guide to creating, using, and maintaining open data portals. Safe Software. 2021. Available from:

https://engage.safe.com/blog/2021/04/guide-creating-using-maintaining-open-data-portals/

[14] Cesar A. What's wrong with open-data sites–and how we can fix them. Scientific American. 2016. Available from: https://blogs.scientificamerican.com/guest-blog/what-s-wrong-with-open-data-sites-and-how-we-can-fix-them/

Chapter 3

Perspective Chapter: Open-Source Scientific Software and Research Data in the Fourth Paradigm of the Sciences and Digital Humanities

Alexandre Ribas Semeler, Edson Mário Gavron, Adilson Luiz Pinto and Fabio Lorensi do Canto

Abstract

Data-driven sciences opened a new dimension in science and digital humanities, beginning a revolution in scientific thinking. In this context, a chapter aims to demonstrate datafication in this field. Computer software mediates data enhancement to develop scientific investigation. The readers are the scientific community in general and Library and Information Science professionals. The landscape of open-source scientific software like research data repositories, including DSpace, EPrints, Fedora, Dataverse, CKAN, dLibra, and eSciDoc, is approached. The sole purpose is to help readers understand how the complexity of datafication can serve as the basis for identifying the theoretical issues relevant to work investigation activities with scientific software and research data. The chapter is about the emergence of new fourth paradigms transforming the research world of science and digital humanities; the fourth paradigm is a concept that refers to a new way of doing science. The other three paradigms are empirical, experimental, theoretical, and simulation enhanced. Finally, it can be concluded whether information about open-source software to develop research data repositories that enable access and preservation of a wide range of research data types.

Keywords: data-driven sciences, fourth paradigm, research data, repository software, DSpace, EPrints, Fedora, Dataverse, CKAN, dLibra, eSciDoc

1. Introduction

Data are now stored in ever-available conditions and can be globally accessed from anywhere by any user. Digital data are a new form of information generated by all human activities with digital technologies. The data landscape manifests a strong tendency toward study and new practices that a librarian, archivists, and other Library and Information professionals in the software era. Data-Driven sciences and digital humanities are surrounded by global collaboration and the new data-information technology infrastructures used by the various branches of science and digital humanities. Scientists and social scientists' professionals need network-computing resources to integrate, federate, and analyze data information in different locations and times [1].

In this context, the chapter is organized into the following topics. First, demonstrate a datafication of science and digital humanities research. Second, the concept of research data and scientific software; and following, present the type of the foremost open-source software for developing data repositories.

We live in a datafication of society; according to Ref. [2], communities, organizations, and all people live in a time when data is collected on anything, anytime, anywhere. A datafication of science and digital humanities research is a basis of the data-driven paradigm, a consolidation of the fourth paradigm, which we understand here as an adjective that qualifies data-oriented processes in scientific investigations. Data-driven sciences and digital humanities opened a new dimension to scientists, the data and software era universe, causing a revolution in scientific thinking. The devices, software, and hardware created through technical mediation transform our experience and raise relevant questions, creating a basis for studies in data technology. In this context, technology not only enhances our capacity to be in the world, but its impact also changes fundamental branches of the theory of knowledge, such as metaphysics, epistemology, ethics, politics, science, and other conventional ways of looking at the natural world.

In summary, the chapter is about the fourth paradigm, the emergence of new paradigms transforming the world of science and digital humanities using research data and scientific software. The fourth paradigm, a concept explained in Refs. [3–6], refers to a new way of doing science. Digital and electronic science. It is the crossroads between technology and scientists. The other three paradigms are empirical, experimental, theoretical, and simulation enhanced. It is believed that digital technologies have revolutionized scientific methods. The conventional methods comprise empirical observation/exploration, theorization, and simulation [3].

In this point of view, the product of the fourth paradigm is digital research data; this digital science product is complex and fluid and includes the scientific recorded information; it is necessary to support or validate research project observations, findings, or outputs of all science and digital humanities.

Research data is collected, observed, or created in digital form for analysis to produce original research results in science and digital humanities. Does scientific evidence use a digital file, irrespective of its content or form (e.g., in print, physical objects, or other forms are digitalized), that comprises research observations, findings, or outcomes, including primary materials and analyzed data? Virtually all types of digital information have the potential to be research data if they are being used as a primary resource for scientific investigations [7, 8].

Computer science brought a new scientific paradigm that modified how scientific investigations was conducted. Technologies created new enquiry possibilities (or types). Experimental data, collected through instruments or generated by simulation, were processed by complex software systems, and only then was the resulting information (or knowledge) stored in computers.

Scientists analyzed the data only at the processing end. This context signified an essential change in the process of scientific thinking, which was replacing hypothesis formulation, experimentation, and results analysis with hypothesis formulation and answer search in the databases.

From this point of view, this chapter explains that the input of the scientific software is a digital research data; this data results from any systematic investigation involving observation, scientific experimentation, or scientific simulation. Digital research data [9] depends on the domain or scientific discipline and may differ in investigative methodologies, it must be identifiable, citable, visible, recoverable, interpretable, and reusable; thus, the requirements of consistency and precedency must be considered.

Scientific software can be understood as a set of rules or patterns of meaning and relationships, similar to political rules or scientific principles, which are systematically developed. Scientific software generates research data in digital form. Scientific software is [7, 8, 10] used by scientists during their formal education and training of new scientists to create and learn scientific technologies. Scientific software plays an essential role in decision-making, for example, making weather predictions based on climate models and computing evidence for research publications.

The other perspective in this chapter is an open-source software research data repository like scientific software. This technical and digital organizational system information helps researchers manage and store data. Also, it eases the search for and access to research data in one or multiple sources, both internal and external, in the repository. Digital repositories appeared in the early 1990s to disseminate publications and other types of digital objects. Developed with free, open-source software, they relate to openness movements. The knowledge includes Open Source Initiatives, Open Access, and Open Data. One of the first initiatives to create repositories was the arXiv developed at Cornell University in the United States, which started in 1991 as a digital library for preprints in Physics [7, 8].

This chapter discusses the importance of data-driven sciences and digital humanities. The role of information technology infrastructures in global collaboration among scientists. It explores how data-driven sciences and digital humanities trust networked computing capabilities to integrate, federate, and analyze information across different locations.

The chapter also addresses the change in traditional scientific research models by introducing digital data and scientific software, highlighting the status of research data repositories as systems for organizing and accessing scientific information records.

2. Datafication of science and a digital humanities

Digital data technologies transformed the way scientific research is conducted, leading to new scientific methodologies. This revolution is not limited to the natural sciences; the impact of digital data technologies on scientific methods can change how we approach research across all fields of study as humanities.

Data as a result of research investigations is research data in cyberspace are virtual and show characteristics of an independent world; these characteristics can be similar to or different from those found in data generated to represent the natural world. In this sense, two main components define data studies. The first is the study of the standards and norms that define the data itself. The purpose of this component is to explore the nature of data and related scientific issues without considering the meaning of the data in the natural world. The second component is the study of the rules of the natural world, as reflected by the data [11].

Data is always-on and can and will be considered as any object created digitally (digital-born) or converted to digital form (digitized), which can be used to generate insights into scientific knowledge. Data is a product of scientific investigations and an input for research. Data can be considered electronic files containing information collected systematically, structured, and documented to serve as input for further scientific studies.

Research data is scientific records information. Data is the raw research material produced through any systematic information collection for analysis [12, 13]. Research data is the same, but in hard science and digital humanity, events, and evidence can be recorded, collected, observed, and generated for scientific investigation analysis and may produce research results for a specific scientific study.

Research data must be able to be collected systematically, structured, and documented to serve as input for further research. Research data can be characterized in many ways according to its nature, origin, or status in the scientific investigation workflow. Research data may differ in its typologies. Depending on the subject or scientific discipline, the definitions may include a broad typology of digital and non-digital objects [13]. Research data proliferate because of the impulse innovations of information technologies, as technology is one of the primary data-generating sources. Thus, it is evident that research data will differ according to scientific methods and that data depend on the specific characteristics of each scientific discipline. In this sense, the next topic attempts to delimit the definition of scientific software by its usage in different scientific disciplines. The complexity of the research data concept reveals that the convolution of scientific software to understand the various dimensions between research data in sciences and digital humanities needs scientific software usage.

3. Scientific software

Scientific software can be conceptualized in multiple ways. In summary, this chapter tries a metaphor for the general concept of technology [5–8, 10, 14–17]. It can be seen as a technical process or methodology applied in such as computers or smartphones, which are concrete objects that require specific skills and training to use effectively. Additionally, scientific software can be understood as a set of rules or patterns of meaning and relationships, similar to political rules or scientific principles, which are systematically developed.

Computational simulation and data-intensive science revels a techno-epistemology in digital science and digital humanities and materialize scientific investigation methods software; this introduction of technologies in all methodologies change as a developed result of using digital science and digital humanities, the scientific software is part of a cyberinfrastructure, as conventional scientific models, such as theoretical and empirical models, no longer support themselves. Ins point of view, does it no longer apply? Or allow does it test theories through simulation? It emphasizes [14] the role of technology as a rule or methodology. Lastly, scientific software can be viewed as a system where hardware and software interact and are considered in the context of their users [17].

This perspective recognizes the impact of technology on extending human capacities and shaping our understanding of the world. The social construction theory suggests that technology and society are interconnected and mutually shape each other. It highlights the complex nature of technological systems and emphasizes that technology is not neutral, independent, or autonomous but a product of human actions and societal influences [14, 17, 18].

Thus, one proposal of this chapter will highlight the knowledge about scientific software used for collecting, manipulating, analyzing, and visualizing research data, called research data repositories. Thus, we assume that some of these software's are required to preserve and curate all research. Research data repositories are used to provide research data access and preservation.

4. Research data repositories

Research data repositories are part of a cyberinfrastructure of the fourth paradigm that Library and Information professionals must master. Thus, one proposal of this chapter will highlight the knowledge and skills necessary for collecting, manipulating, analyzing, and visualizing available data in research data repositories. Hence, we assume that librarians require some of these skills.

The different kinds of technological systems created to support research data compose a rich universe of digital information that register the knowledge resulting from scientific investigation. The research data repositories may be available from two classes of providers: data providers and service providers. Data providers maintain digital document repositories and implement protocols such as the Open Archives Initiative Protocol for Metadata Harvesting (OAI-PMH) to make their networked metadata available. Service providers collect data for building value-added services for data, offering metadata searches or other services. The data repository should be as interoperable as possible [19–21].

The global distribution of research data repositories is cataloged by re3data, the international research data repositories registry. The re3data repository covers different academic subjects and has been registered since 2012 with approximately 3000 in international scenarios [22]. It relates data repositories for permanent storage and provides access to research datasets for funding bodies, editors, and academic institutions to promote a culture of sharing, access, and visibility for research data [23].

As said by Ref. [24], the essential function of trusted digital repositories is "[...] a mission to provide reliable, long-term access to managed digital data resources to its designated community, now and into the future." The digital repositories must accept responsibility for the long-term maintenance of digital resources and research data, which implies having a system that supports the interoperability of digital information, provides physical responsibility and sustainability in digital media, and is designed to ensure the management, access, and security of the digital objects deposited in it. Data Repositories may be implemented with a diversity of software technologies. According to Ref. [23], the type of software applications for development repositories is CKAN (97), DSpace (126), Dataverse (167), DigitalCommons (5), EPrints (34), Fedora (48), MySQL (90) Nesstar (19), Opus (3), dLibra (3), eSciDoc (5), other (649), unknown (1187).

5. Main open-source software for the development of data repositories

A repository should establish evaluation methodologies, performance policies, and practices that can be audited and measured, such as sustainability, security, and technology infrastructure rules. Repositories should commit to reliability standards, such as those of the OAIS, which resulted in the standard ISO16363:2012, which lists the essential criteria for trusted digital repositories [19].

Data repositories may be implemented with a diversity of software technologies. However, generally, they are elaborated with free, open-source software platforms

Software	Description
DSpace[1]	It was released by the MIT Libraries and Hewlett-Packard Labs in 2002 to provide a repository system for digital documents resulting from research or intended for education and distributed with an open-source license. In 2007, MIT and HP created the DSpace Foundation, a non-profit organization, to promote the platform and support its users. In 2009, this support went to the DuraSpace Foundation, a non-profit organization dedicated to open-source and cloud technologies for libraries, universities, research centres, and cultural heritage organizations.
EPrints[2]	The EPrints Repository Software is maintained by the School of Electronics and Computer Science, University of Southampton, UK. The platform is distributed based on an open-source license. In addition to offering the standard functionalities of institutional repositories, EPrints Services has been associated with a consulting team that can follow a project to install a repository from the analysis and customize development to provide management services.
Fedora[3]	Fedora is not a platform for repositories like DSpace or EPrints but an extensible architecture that can be used to develop software for repositories. Created by Cornell University, it is currently maintained by the DuraSpace Foundation. It has principles such as aggregating local content and distributed digital objects and associating these with services [Fedora]. The architecture also includes a relationship model based on the W3C's RDF, used to bind objects to their components. It is available with an open-source license and has been used in many applications for digital libraries, archives, institutional repositories, and learning object systems.
Dataverse[4]	The Dataverse Network (DVN) is an open-source software used to manage data collection. The main goal of DVN is to solve data-sharing problems and replication of scientific information on the web. It supports archiving, backup, information retrieval, persistent identifiers based on fixed data patterns, metadata conversion, and preservation. The DVN facilitates the creation of the so-called dataverse. A dataverse can be a web archive or repository to store and share scientific data. The development of DVN software began in 2006 at the Institute of Quantitative Social Sciences (IQSS) at Harvard University. The concept governing the implementation of DVN is data replication; that is, a dataverse must contain the information necessary to reproduce an original study to provide an empirical analysis of the exact process of how the research data was generated or produced.
CKAN[5]	CKAN is a tool for creating open data repositories. It is used to manage and publish collections of data. It is widely used by research institutions and other organizations that collect data. CKAN is open-source software with an active community of developers who develop and maintain a growing library of CKAN extensions.
dLibra[6]	dLibra is the first Polish system for building digital libraries and has been developed by the Poznan Supercomputing And Networking Center (PSNC) since 1999. dLibra is a digital library research tool used at the PSNC since 1996. The dLibra system is now the most popular software of this type in Poland. dLibra enables the building of professional repositories of digital documents that external individuals and systems can access on the internet. Communication and data exchange is based on well-known standards and protocols, such as RSS, RDF, MARC, DublinCore, and OAI-PMH.
eSciDoc[7]	eSciDoc is an e-Research environment explicitly developed by scientific and scholarly communities to collaborate globally and interdisciplinarily. It comprises core functionality, including a Fedora repository (eSciDoc Infrastructure), a set of complementing services (eSciDoc Services), and an application built on top of the infrastructure and the services (eSciDoc Applications) that enables innovative e-research scenarios. Scientists, librarians, and software developers can work with research data, create novel publications, and establish new scientific and scholarly communication methods. The software is available as open-source software. The development of the eSciDoc Infrastructure ended in 2012. We do not recommend using the software for new projects due to security issues. Some former eSciDoc applications are still under active development. Please contact the Max Planck Digital Library to learn more about reuse options for the software.

[1]*Ref. [25].*
[2]*Ref. [26].*
[3]*Ref. [27].*
[4]*Ref. [28].*
[5]*Ref. [29].*
[6]*Ref. [30].*
[7]*Ref. [31].*

Table 1.
Open-source software for the implementation of data repositories.

already known and internationally implemented by the library community, such as DSpace, E-prints, Fedora, Dataverse, CKAN, dLibra, and eSciDoc. This software was developed to collect, preserve, and disclose research data publications, but they can aggregate any content in a digital format with **Table 1**.

The relevance of this software in **Table 1** represents the packages because they facilitate access to, interoperability, and preservation of digital research data. The diverse ways research data flows in these systems, mainly on the internet, accentuates the need to organize, comprehend, preserve, and analyze information and knowledge that can be extracted from this digital information medium.

A research data repository is a technical information and digital organizational system to help researchers with data management and storage and ease of searching for and accessing research data in one or multiple internal and external sources in the repository. Data repositories are essential to the research cyberinfrastructure intended for preservation, long-term access, and reutilization [19].

6. Conclusions

The traditional models of scientific investigation changed with digital research data and scientific software. The fourth paradigm is a field of knowledge that also focuses on translating scientific methods to computers. As a paradigm inclusive of technology, the fourth paradigm reveals new interfaces to all scientific domains, the computer, the world wide web, and the data landscape of all digital science and digital humanities activities. The essential characteristics of the fourth paradigm are digitally-enhanced aspects, how scientists analyze, manage, gain access to, and share digital data through scientific software. Digital scientists utilize networked data and materials to formulate new information through cross-comparison and manipulation. Therefore, the fourth paradigm thrives when datasets are shared and accessible.

Data manipulation achieves paradigm exploration needs scientific software; thus, datasets used as the primary form of experimentation need scientific software. Researchers can find patterns and develop new inquiries across disciplines by manipulating and cross-comparing datasets, which requires scientific software.

The digital software in the sciences amplifies itself when mediated by computers, mainly if networks like the web are included digital research data. Currently, research data are stored in always-on conditions and can be accessed globally at any time by any user. The exponential growth of data generation is related to everything we use during our daily routine. Technological systems to support research data have created a vast universe of scientific software, the knowledge resulting from scientific investigation, and the uses of scientific software. The diverse ways research data flows in these systems, mainly on the internet, accentuates the need to organize, comprehend, preserve, and analyze information and knowledge that can be extracted from this digital information medium.

According to the context, a consolidated environment of digital research data, scientific software is the basis for developing research data repositories. Research data repositories are registered in the most diverse fields of knowledge and distributed globally. However, who is a humanities or hard sciences professional scientist in the fourth paradigm? In the software era, a Library and Information professional manages and organizes digital research data with software such as DSpace, E-prints, Fedora, Dataverse, CKAN, dLibra, and eSciDoc. With the advent of these software tools, the fourth paradigm consolidates the data share era of the cyberinfrastructure, database systems, and information management platforms to handle and provide access to ample data assets effectively.

Based on the provided about scientific software and research data, it could be further explored: the impact of data-driven sciences and digital humanities: how these technologies have transformed research practices and outcomes.

The ethical and privacy considerations in data-driven sciences: discussing informed consent, data anonymization, data security, and responsible data use to contribute to a more well-rounded exploration of data-driven sciences. The emerging trends and future directions on the current state of data-driven sciences and digital humanities are exploring emerging technologies, such as artificial intelligence, machine learning, and big data analytics, and their potential impact on research practices would provide insights into the future of data-driven sciences.

Author details

Alexandre Ribas Semeler[1*], Edson Mário Gavron[2], Adilson Luiz Pinto[2] and Fabio Lorensi do Canto[2]

1 Federal Univeresity of the Rio Grande do Sul, Geosciences Institut, Porto Alegre, Rio Grande do Sul, Brazil

2 Federal Univeresity of Santa Catarina, PGCIN, Florianópolis, Santa Catarina, Brazil

*Address all correspondence to: alexandre.semeler@ufrgs.br

References

[1] Hey T, Hey J. Fourth paradigm and its implications for the library community. Library HiTech. 2006;**24**(4):515-528. Available from: http://eprints.rclis.org/9202/ [Accessed: May 10, 2023]

[2] Van der Aalst WMP. Data scientist: The engineer of the future. In Mertins K, Bénaben F, Poler R, Bourrières J-P, editors. Enterprise Interoperability VI, Interoperability for Agility, Resilience and Plasticity of Collaborations (Proceedings of I-ESA 2014, Albi, France, March 24-28, 2014). Proceedings of the I-ESA Conferences. Cham: Springer; 2014. p. 13-26. DOI: 10.1007/978-3-319-04948-9_2 [Accessed: May 10, 2023]

[3] Tansley HS, Tolle K, editors. Fourth Paradigm: Data-Intensive Scientific Discovery. Microsoft; 2009. Available from: http://research.microsoft.com/en-us/collaboration/fourthparadigm/ [Accessed: May 10, 2023]

[4] Floridi L. What is the philosophy of information? Metaphilosophy. 2002;**33**(1-2):123-145

[5] Floridi L. The Philosophy of Information. Oxford: Oxford University Press; 2010

[6] Floridi L. Steps forward in the philosophy of information. Etica & Politica/Ethics & Politics. 2012;**14**(1):304-310. Available from: http://www2.units.it/etica/2012_1/FLORIDI.pdf [Accessed: May 10, 2023]

[7] Kanewala U, Bieman JM. Testing scientific software: A systematic literature review. Information and Software Technology. 2014;**56**(10):1219-1232. DOI: 10.1016/j.infsof.2014.05.006 [Accessed: May 10, 2023]

[8] Rice R, Southall S. The Data librarian's Handbook. London: Facet Publishing; 2016

[9] Shcmillen H. Library and Information Science Education and eScience: E Current State of ALA Accredited MLS/MLIS Programs in Preparing Librarians and Information Professionals for eScience Needs. Denver: Capstone Projects, Paper 1. Available from: http://digitalcommons.du.edu/lis_capstone/1; 2015 [Accessed: May 10, 2023]

[10] Hannay JE, Langtangen HP, Macleod C, Pfahl D, Singer J, Wilson G. How do scientists develop and use scientific software? 2009 ICSE Workshop on Software Engineering for Computational Science and Engineering. 2009:1-8. DOI: 10.1109/SECSE.2009.5069155 [Accessed: May 10, 2023]

[11] Zhu Y, Xiong Y. Towards data science. Data Science Journal. 2015;**14**:8. DOI: 10.5334/dsj-2015-008 [Accessed: May 10, 2023]

[12] Kellam L, Thompson K. Introduction to Databrarianship: The Academic Data Librarian in Theory and Practice. Chicago: Association of College and Research Library; 2016

[13] Henderson, M. Data Management: A practical guide for Librarians. Lanham: Rowman & Littlefield Publishers; 2017

[14] Kline J. What is technology? In: Dusek V, editor. Philosophy the Technology: The Technological Condition an Anthology. Malden: Blackwell Publishing; 2006

[15] Dusek V. Philosophy the Technology: The Technological Condition an Anthology. Malden: Blackwell Publishing; 2006

[16] Cupani A. Filosofia da Tecnologia: Um convite. Florianópolis: Ed. Da UFSC; 2013

[17] Semeler AR, Pinto AL, Vianna WB. E-science: An epistemological analysis based on the philosophy of technology. IFLA Journal. 2017;**43**(2):198-209. DOI: 10.1177/0340035216678235 [Accessed: May 10, 2023]

[18] Vallverdú J. Computational epistemology and e-science: A new way to thinking. Minds and Machines. 2009;**19**(4):557. Available from: https://www.academia.edu/493057/Computational_Epistemology_and_e-Science_A_New_Way_of_Thinking?auto=download [Accessed: May 10, 2023]

[19] Kindling M, Pampel H. Informations in frastrukturangebote für digitale Forschungsdaten. E(hren) Journal. 2017;**2017**:15-33. DOI: 10.18452/2341 [Accessed: May 10, 2023]

[20] OPEN Archives INITIATIVE. Protocol for Metadata Harvesting (OAI-PMH). Available from: http://www.openarchives.org/pmh [Accessed: May 10, 2023]

[21] Garcia PA, Sunye M. O Protocolo OAI-PMH para Interoperabilidade em Bibliotecas Digitais. CONGED. Available from: http://conged.deinfo.uepg.br/~iconged/Artigos/artigo_09.pdf; 2006 [Accessed: May 10, 2023]

[22] Semeler A. Re3data scripts to parsing, scraping, and visualization (beta). Zenodo. 2023. DOI: 10.5281/zenodo.7956947 [Accessed: May 10, 2023]

[23] RE3DATA. Available from: http://www.re3data.org/about [Accessed: May 10, 2023]

[24] Organization for Economic Co-operation and Development (OECD). OECD Principles and Guidelines for Access to Research Data from Public Funding. Australia: Organization for Economic Co-operation and Development; 2004. Available from: https://www.oecd-ilibrary.org/docserver/9789264034020-en-fr.pdf?expires=1702564653&id=id&accname=ocid54025470&checksum=8557FD16AD7C853110A905D2628E4693 [Accessed: May 10, 2023]

[25] DSPACE. Available from: www.dspace.org [Accessed: May 10, 2023]

[26] E-PRINTS. Available from: http://www.eprints.org/uk/ [Accessed: May 10, 2023]

[27] FEDORA. Available from: https://getfedora.org/pt_BR/ [Accessed: May 10, 2023]

[28] DATAVERSE. Available from: https://dataverse.org [Accessed: May 10, 2023]

[29] CKAN. Available from: https://ckan.org / [Accessed: May 10, 2023]

[30] DLIBRA. Available from: http://kpbc.umk.pl/dlibra/help?id=about-dlibra [Accessed: May 10, 2023]

[31] ESCIDOC. Available from: https://www.escidoc.org [Accessed: May 10, 2023]

Chapter 4

Building a Big Data Platform Using Software without Licence Costs

Vassil Vassilev, Viktor Sowinski-Mydlarz, Pawel Gasiorowski, Sorin Radu, Sabin Nakarmi, Martin Hristev, Reza Baghaeishiva and Tarun Bali

Abstract

This chapter presents the experience in developing and utilizing Big Data platforms using software without license costs, acquired while working on several projects at two research institutions – the Cyber Security Research Centre of London Metropolitan University in the United Kingdom and the GATE Institute of Sofia University in Bulgaria. Unlike the universal computational infrastructures available from large cloud service providers such as Amazon, Google, Microsoft and others, which provide only a wide range of universal tools, we implemented a more specialized solution for Big Data processing on a private cloud, tailored to the needs of academic institutions, public organizations and smaller enterprises which cannot afford high running costs, or do significant in-house development. Since most of the currently available commercial platforms for Big Data are based on open-source software, such a solution is fully compatible with enterprise solutions from leading vendors like Cloudera, HP, IBM, Oracle and others. Although such an approach may be considered less reliable due to the limited support, it also has many advantages, making it attractive for small institutions with limited budgets, research institutions working on innovative solutions and software houses developing new platforms and applications. It can be implemented entirely on the premises, avoiding cloud service costs and can be tailored to meet the specific needs of the organizations. At the same time, it retains the opportunity for scaling up and migrating the developed solutions as the situations evolve.

Keywords: big data, AI, data platform, private cloud, public domain

1. Introduction

According to the recent Gartner report on strategic technology trends [1]. Platform Engineering is one of today's top 10 trends influencing enterprise strategies. *Data Platforms* deal with the entire lifeline of digital data from the moment it is generated at the data source through the communication channels which collect it and transport it to the destinations where it is accumulated, transformed, stored and processed, all the way to the actual interpretation of the result at the destination [2]. Behind its continuing expansion are two orthogonal but interconnected factors of development in the digital age – the evolution of computing technologies and the vast amount of digital data available from different sources which are potentially useful for different purposes.

The evolution of data processing in the digital age went through four main paradigms, constantly vacillating between centralization and decentralization:

- Localization of the data processing in a single physical place (*desktop*, *embedded* or *mobile* device). Dominant during the early computing and communication devices.

- Distribution of the processing power across multiple physical locations on a computer or communication network (*client-server* or *peer-to-peer* architecture). Exploits the opportunities for sharing provided by contemporary digital networking on local, regional or global scale.

- Concentration of the processing power by creating virtual locations for hosting multiple processors (*cloud* or *edge* computing). Initially supported and later even enforced by the big vendors because it facilitates the performance of computations on a scale previously not imaginable and unaffordable.

- Virtualization of the computation by creating logical processors in different physical and virtual locations (*blockchains* or *data spaces*).

Since the beginning of the new Millenium this conceptual evolution was accelerated by the evolution of software technologies. Firstly, two complementing enabling technologies contributed to this evolution: *virtualization* of the computational infrastructure and *containerization* of the execution environment (see **Figure 1**). While the virtualization allows applications to be executed in the environment of a virtual operating system, chosen by convenience, containerization allows the execution to be identical, regardless of the deployment location. The contemporary cloud provision for deployment employs both approaches, thus supporting the high scalability of the hardware and software infrastructure as well as the mobility of the applications across the platform locations. The transparency of the application deployment in such a case is provided by the container management system, which is a key element of each cloud provision [3].

This shift of the computational paradigms is paired with the expansion of the software systems towards service-oriented architectures (SOA), where the applications are not executed in isolation but as an element of a *service workflow* (see **Figure 2**). The platforms with SOA architecture can *orchestrate* the services, thus adding the possibility for automation of workflow planning and execution.

Unlike cloud computing, which essentially pushes for centralization of data processing on the cloud, the SOA better supports distributed architectures for peer-to-peer data processing in blockchains and data spaces. However, the two architectures are not antagonistic, since the service orchestration is also popular on the data platforms, while they may also act as data service providers or data service consumers in data spaces [4].

What is extremely encouraging is that all these opportunities are freely available and can be implemented in-house, tailored to the needs of the organization. This chapter presents the experience in implementing data platforms on a private cloud for

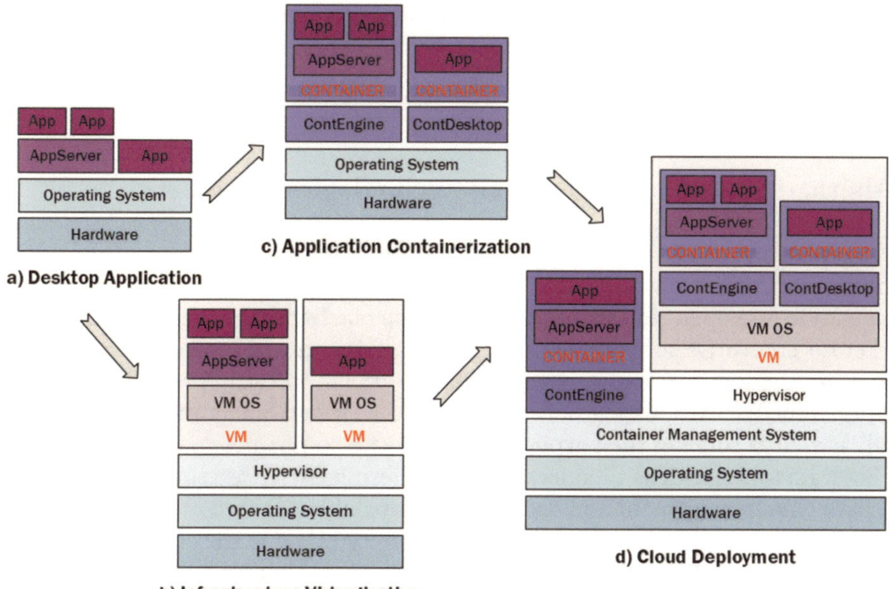

Figure 1.
Evolution of data processing from the desktop to the cloud.

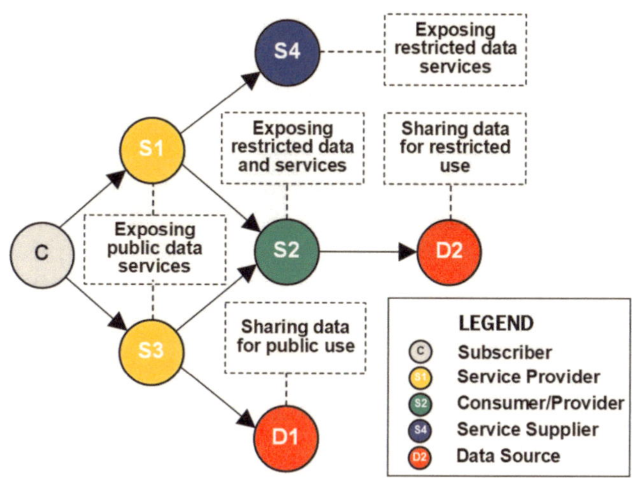

Figure 2.
Data processing workflow in a service-oriented architecture.

processing Big Data using software without license costs, which can be obtained from the public domain, or as community editions of commercial products.

The plan of the chapter is as follows. In the next section, following a brief review of some of the existing solutions, we will discuss the main alternatives which drive the design of data platforms. In the subsequent section, we will describe the main components of the platform. After that, we will present several pilot projects, which we implemented using such platforms – one for real-time security analytics, one for outdoor air quality monitoring and one for more complex urban development combining both outdoor and indoor environment factor analysis. At the end we will discuss the lessons learned, will formulate some recommendations and will discuss some future enhancements of the platform.

2. Alternatives and choices: models, technologies and tools

Data platforms and, specifically, Big Data platforms are complex systems which combine powerful hardware and software. They target decision-making, business organization and operation management. On the one hand, the main players in the market for enterprise software offer their own tools and whole ecosystems for data processing and Big Data management. Both traditional software powerhouses, such as IBM, Hewlett-Packard, Oracle and Amazon in the United States and SAP in Europe [5–9], as well as some of the purpose-built software companies specialized in marketing Big Data tools, such as Cloudera [10] are offering enterprise suites with extremely powerful data management and data analysis capabilities, based on well-established concepts originating in the open-source community [11]. On the other hand, the global service providers, such as Amazon, Google, and Microsoft [12–14] are hosting most of these tools on their own cloud premises thanks to the technologies of virtualization, containerization and orchestration which are the foundation of cloud computing. In some cases, this symbiosis goes even further by embedding mechanisms for utilization of the specific storage infrastructure, like the recent Cloudera platform **CDH**, which is seamlessly integrated with the object storage of **AWS** [12]. However, although midsize companies and smaller software houses prefer to rely on software vendors and service suppliers, the price tag associated with it is exorbitantly high. In most cases, it is out of reach for many private and public organizations, and they typically look for a bespoke solution on their own premises instead. An additional advantage of having custom-built data platforms on the premises is the compatibility and the opportunity for subsequent migration to an enterprise platform. Since most of the current data platforms are assembled out of software products and systems which originate in the open-source community, in the case of scaling up these in-house platforms, it would be possible to migrate the applications much more easily. This section will systematically consider the necessary decisions to prepare ourselves for building a custom-tailored in-house platform for processing Big Data on a private cloud. This would be a competitive solution for data-intensive but relatively small organizations such as research institutes, local councils, public agencies and SMEs.

2.1 Conceptualization of data processing, workflows and data services

To build a platform which supports the entire lifecycle of the data from the moment it has been generated until the moment the result of its processing is interpreted, we need to have a clear conceptual understanding of the use of such a platform for its intended use. The most economical and useful way to do it is to have a model, presented graphically using a set of diagrams, which is complemented by a description of the typical business scenarios and detailed technical specifications:

- Data: data sources, data formats and data pre- and post-processors.

- Platform infrastructure: hardware and software components and subsystems.

- Data processing architecture: software pipelines and data flows.
- System interaction: communication ports, protocols and operations.
- Operation timelines: scenarios for using the platform for specific business tasks.

From a software engineering perspective, the standard way of doing this is using UML as a modeling language. It provides the most complete set of diagrams, which makes the design informative and technically sound, but it may also lead to "analysis paralysis" in complex tasks such as the platform design. Because of this, we prefer a logical approach, which combines the modeling of the conceptual ontologies with flowcharts of their use. It also guarantees consistency across the diagrams but is more economical than UML. As an example, **Figure 3** shows the ontology of data processing. Some of the concepts in it are physical and will later become software components of the platform, while others are purely logical, helping to understand the problem, conceptualize the processes and explain the results of processing the data on the platform. In ontology, we can model the data processing operations as relations between the states and organize them into a hierarchy, similar to the concept taxonomy. The full ontology of the platform, as we modeled it, includes more than 600 different concepts, but the development of the platform can be incremental so that the ontology can be a useful starting point for the design. As a step towards this, **Figure 4** illustrates the sharing of both data and services for processing it across several business projects analyzing the environment impact on urban life. Three projects are considered "first level": monitoring outdoor air pollution (Project 3), analyzing the medication prescriptions in the local pharmacies (Project 2) and mapping them in the local

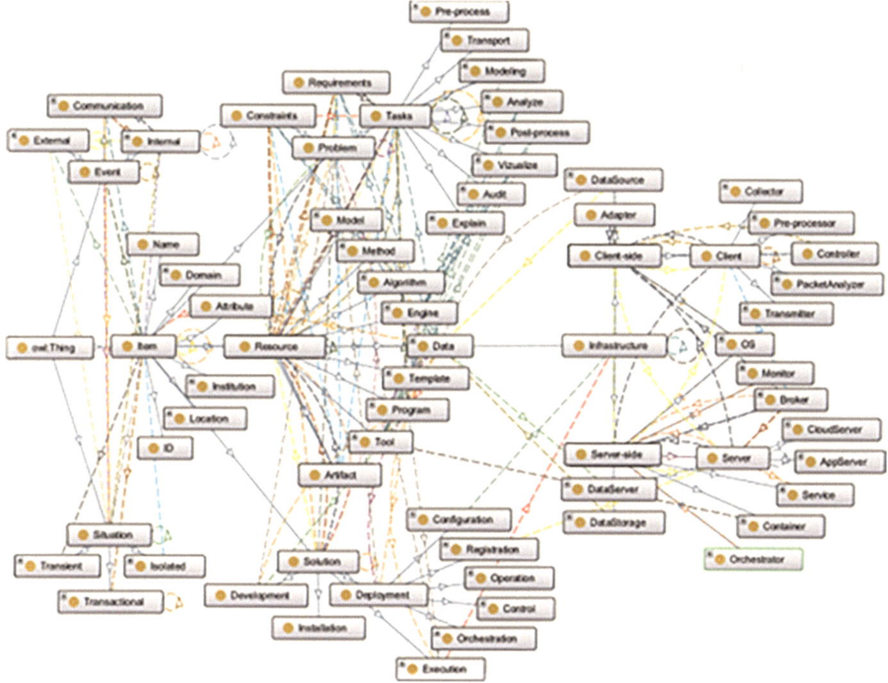

Figure 3.
Ontology of data processing on the data platform (core fragment).

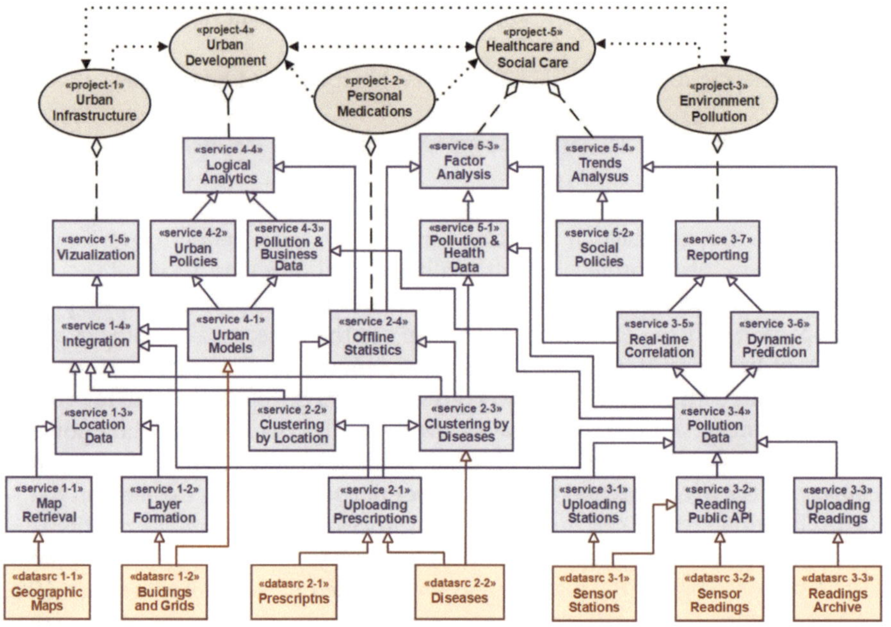

Figure 4.
Data platform as a service-oriented architecture (urban life projects).

area using public sources of geolocation information (Project 1). Additionally, there are two cross-domain projects for secondary analysis of the impact of environmental pollution on public healthcare (Project 5) and on urban development (Project 4), which use data, existing services, and some of the results produced by the "first-level" projects. They can provide deeper, policy-driving analysis of the environmental impact based on factors, trends and heuristic methods. The diagram shows the data sources, data flows and dependencies between the services. It is sufficient for the design of software architecture according to both the privacy and the sharing policies.

The conceptual model must lead to a practical implementation of the scenarios of use and the most natural way to do this is to map it to a service-oriented architecture (SOA). All aspects of the SOA – functional, temporal, causal, etc., can be designed so that they retain semantic consistency with the conceptual model. The integrity can be guaranteed by stereotyping the ontological entities, which preserves the meaning across different diagrams, similar to the stereotyping in UML. For our example, **Figure 5** shows the workflow of correlating data in the form of a directed graph, which implements the "Real-time Correlation" service from **Figure 4**.

We have found this approach more useful than the use of UML at the early stage of design of the platform since it ignores low-level details and combines many different aspects of the model in a single diagram, which facilitates early grasping of the essentials and organizing the work on a technical level in an optimal way. The next sections will go systematically through the alternative choices and detailed descriptions of the tools which map this conceptual model to a working system.

2.2 Exploring data and metadata

The enterprise data platforms, as well as the platforms built on the public cloud, have configurable components from a chosen software *ecosystem*. Their universality is

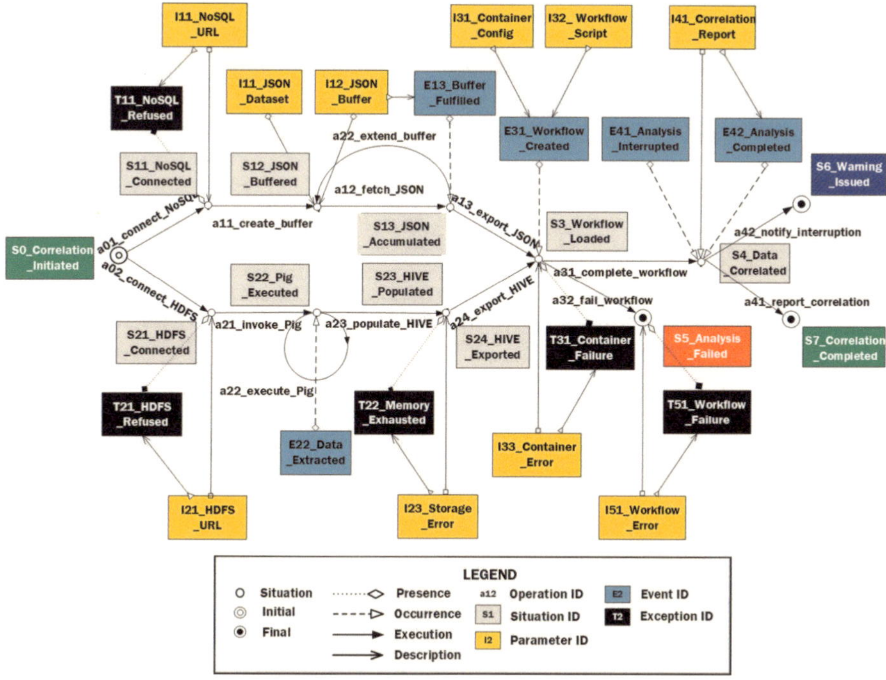

Figure 5.
Workflow for data processing on the platform (data correlation scenario).

restricted only by the software vendor, or by the support provided by the public cloud to the installed ecosystem. The private cloud-based platforms do not have these limitations, although their scope is more limited. This allows custom-tailored design, which better accounts for the specific data to be processed on the platform. Quite a few characteristics of the data need to be considered at this stage:

- The data differs significantly in a variety of ways – formats, granularity, volume, noise, location, etc.

- Data processing is performed along a complex workflow of operation – sampling, aggregation, buffering, feature selection, training, validation, analyzing, merging, interpretation, explanation, etc.

- The tasks for data analysis have large diversity – detection, recognition, classification, correlation, factorization, prediction, etc.

- For each analytical task there is a whole variety of methods with different applicability – temporal, structural, logical, model-driven, behavioral, hybrid, etc.

- To reach wider community of users the data processing needs to be comprehensive by providing statistics, reports, explanation, etc.

This variety directly affects the subsequent choices of technologies, methods and tools for building the ecosystem of the platform. The main technical characteristics of

Data types	Sources	Content	Ingestion	Transport
Samples (structured data)	networks, hardware and software	readings, packets, locations	one-off	memory sharing, parameter passing
Artefacts (unstructured data)	media editors, reporters and cameras	documents, images, videos	one-off	FTP/S, HTTP/S, SSH, FTAM, WebDAV, WebSockets
Messages (semistructured data)	messengers, devices and monitors	alerts, logs, messages, emails	one-off	MQTT, AMQP, SMS, IRC, XMPP, ModBus, Websocket, RCS
Streams (fully/ semi-structured sequences)	signal emitters, trackers and video cameras	timeseries, broadcasts, feeds	continuous	HLS, WebRTC, RTSP, RTMP, SRT, MPEGDASH, ModBus
Datasets (fully/ semi/unstructured collections)	spreadsheets, databases and simulators	descriptions, operations, locations, etc.	one-off, batch	FTP/S, HTTP/S, AFTP, OFTP, AS2, WebDAV
Repositories (collections of datasets)	databases, warehouses and data lakes	mixed	batch	supported by the repository

Table 1.
Data characteristics affecting the platform design.

the data which may have an impact on the design choices are shown in **Table 1**. Particular attention is needed for the processing of time series in real-time [15].

Creating data models is one of the early tasks which can be addressed during the design of the platform since it helps both the processing and the persistence of the data. Depending on the data sources, different approaches can be used for modeling: purely relational using ER diagrams, object-relational using UML class diagrams, hypertext using XML or JSON and logical using semantic languages like RDF/RDFS/OWL. An additional important factor to help understanding the data and preparing it for subsequent processing is the *metadata,* the information about the data. It can be used for data cleaning and filtering, for semantic enrichment by adding missing attributes and establishing missing links, as well as for preparing persistent storage and efficient retrieval of the data via semantic indexing.

2.3 Choosing methods for data processing

Data processing on the platform may occur in a variety of contexts and can serve different purposes, so the selection of suitable methods for each task will affect the choice of tools during the development phase:

- Different stages of data pipeline: at the source, before transmission, during transmission, on arrival, before storing, inside the repository, etc.
- Different structure and format of the data: structured (CSV, SQL), semi-structured (JSON, XML, RDF, SVG, etc.) and unstructured (binary, document, graphics, etc.)
- Different preparation of the rough data: filtering, formatting, anonymization, normalization, enrichment, aggregation, buffering, accumulation, etc.

- Different tasks of the processing: detection, classification, recognition, correlation, profiling, prediction, etc.

- Different methods behind the algorithms: statistical, clustering, graph-based, rule-based, model-based, optimization-based, etc.

- Different interpretations of the results: simple reporting, black box explanation, white box explanation, impact factor analysis, etc.

Constructing an enterprise-quality data platform requires well-proven methods for which there are mature tools, guaranteeing robust and reliable operation. In some cases, the data processing may be implemented using publicly available libraries; in other cases, there are no suitable tools, and bespoke software needs to be developed with *Python* as the ultimate language of choice for programming. Although it might be assumed that the more sophisticated methods for data analysis are not easy, with the huge advancement in statistical, behavioural and machine learning methods, their use is a completely feasible task supported with a huge amount of software libraries. Less advanced methods are a bigger problem, since there is rarely a universal and high-quality software available to implement them.

2.4 Software tools for data management and data processing

The technology stack of the platform includes software components for processing the data along its lifecycle, from the sources to the presentation of the results. For most of the necessary tasks, there are software products in the public domain, and many enterprise software products also have community editions, so the composition of a relatively universal data platform using software without license fees is absolutely feasible. **Table 2** contains some software sufficient to compose a technology stack powerful enough for a wide class of typical Big Data scenarios. We used most of them

Operations	Description	Software products
Data Preprocessing	Processing at the source, in transition, before analysis	**Hackolade** (modelling), **Annotator** (annotating), **Amnesia** (anonymizing)
Data ingestion	Transporting files, exports, messages, streams,	**Mosquito** (messages), **Kafka** (streams), **NiFi** (files), **Hop** (general)
Data persistence	Storage in structured, hypertext and binary format	**PostgreSQL** (SQL), **MongoDB** (JSON) **Neo4J** (RDF), **3DCityDB** (CityGML)
Data Postprocessing	Transformation, analysis and integration	**Python** (data-), **Java** (operation-), **JS** (Web-) and shell scripts (OS-centric)
Big data management	Storing, retrieving, mapping & searching Big Data	**Hadoop** (pairs), **Storm** (tuples), **Cassandra** (tables), **HPCC** (clusters)
Big data analysis	Analyzing Big Data using statistical, ML & RL	**Spark** (analysis of distributed data), **Storm** (analysis of streaming data)
Interpretation	Data integration, reporting and visualization	**JS** (texts), **Jupyther** (diagrams), **Cesium** (maps), **Grafana** (general)

Table 2.
Technology stack for data processing on big data platforms.

in several projects, and as we will show in the next section, it is both powerful and highly scalable for working on multiple projects.

A separate consideration is needed for the choice of components for Big Data processing. Our platform is based on Apache Hadoop, which has one of the oldest Big Data ecosystems [16]. It maintains the data in files, which makes it inherently slower. There are some more recent and modern alternatives in the public domain, such as Cassandra, Storm, and HPCC, plus some highly scalable no-SQL databases like Redis, CouchDB and OrientDB [17], which might be more suitable for specific characteristics of the Big Data, like speed of growth, degree of dependency, etc. Our choice of Hadoop was dictated by the need for compatibility with our academic and business partners. Being an originator of many enterprise systems, it may be seen as an "old-fashioned" choice, but Hadoop has a rich ecosystem which is compatible with many commercial systems, so it is still attractive.

The use of additional databases besides the main Big Data repository may seem redundant at first sight but it is justified. Firstly, the rough data, the pre-processed data and the archived data can be used in different projects for different purposes. Keeping the data in a single place in a single format may lead to inefficiency due to the need for conversion. Secondly, sometimes, it is preferable to store specific data in its original format for specific operations, supported natively by specialized database systems. This applies to all non-SQL data – hypertext in JSON format (MongoDB), graphs in RDF format (Neo4J) and 2D/3D in CityGML format (**3DCityDB**).

The separation of preliminary data processing from Big Data processing leads to two different approaches for supporting the analysis of the data. In the Big Data cluster, this can be done using the tools from the Hadoop ecosystem (we use Apache Spark [18]), while in the temporary area of holding the data, it can be processed within the respective databases using their own APIs or externally, using Python.

3. Building the data platform

Our motivation for developing a private cloud-based data platform comes from our interest in the automation of data processing using explicit policy rules for controlling the execution. Initially, we incorporated some of these ideas in the data processing framework for threat intelligence and security analytics of the Cyber Security Research Centre [19]. In another project, we needed computational resources on the public cloud and hosted the software components on Amazon AWS [20]. This experience led us to the concept of implementing a data platform on our own private cloud, which resulted in the solution reported here [21]. In this section, we will describe the main steps of the process.

3.1 Hardware and system software

The starting point for the implementation of the data platform for processing Big Data on a private cloud is the hardware infrastructure. In a realistic scenario, cloud technology requires multiple hosts organized in a virtual cloud infrastructure according to the principle "the more memory – the better". However, being academic institutions primarily focused on research and innovation, both in London and in Sofia, we started with a single server host first. Currently, we are running our platforms on hosts with 128GB RAM/256GB RAM, equipped with 80 TB/120 TB hard disk space (in London/Sofia, respectively). Such capacity might be considered a minimum

Platform role	Description	Software products
Platform management	Register, allocate and control resources	**Nginx** (access control), **Proxmox** (admin, develop, deploy)
Virtual machine management	Emulate different OS on the infrastructure	**Linux KVM** (virtualize)
Container management	Distribute, isolate and synchronize components	**Docker** (containerize), **Kubernetes** (locate), **Zookeeper** (synchronize)
Workflow management	Compose, schedule and integrate workflows	**AirFlow** (orchestrate), **Flask** (integrate)
Process management	Monitor and report the execution	**MLFlow** (monitor, report)
Service management	Search, retrieve and report the consumption	**Elasticsearch** (search, retrieve), **Kibana** (accumulate, report)

Table 3.
System tools for managing of platform resources.

requirement to provide an adequate environment for several teams to work on different projects. Thanks to the scalability of the platform, at a subsequent stage, it can be migrated to a more capable hardware infrastructure with multiple hosts and much larger memory space, as necessary.

Secondly comes the system software, which provides support for the development, deployment and execution of the software components of the platform. This system software is completely independent of the data which the platform is going to process and is neutral to the possible applications deployed to it. The system software which we used for this purpose is shown in **Table 3**.

All system software can be used without license costs and can be obtained from the respective public sites on the Internet. The community versions of enterprise products may have some limitations, but they are still fully functional and can be used for the purpose in non-commercial environment.

3.2 Assembling the platform

Endorsing the previous recommendations leads to the possibility of building an entirely user-empowered platform for processing Big Data on a private cloud, fully compatible with more sophisticated enterprise platforms from the big vendors [5–10]. The software architecture of a similar solution, like our data platform, is shown in **Figure 6**. It is formed out of five different groups of software components:

- Platform management (platform-specific but service and data independent) – management of the platform resources: Kubernetes for container management, Docker for containerization, Proxmox for remote access and resource management, Nginx for security control, Elasticsearch and Kibana for auditing and reporting, Airflow and MLFlow for workflow and process scheduling and monitoring.

- Data ingestion (service specific but data independent) – Mosquitto, Kafka, Nifi and the general-purpose Hop for managing the communication channels, transportation, ingestion and storing.

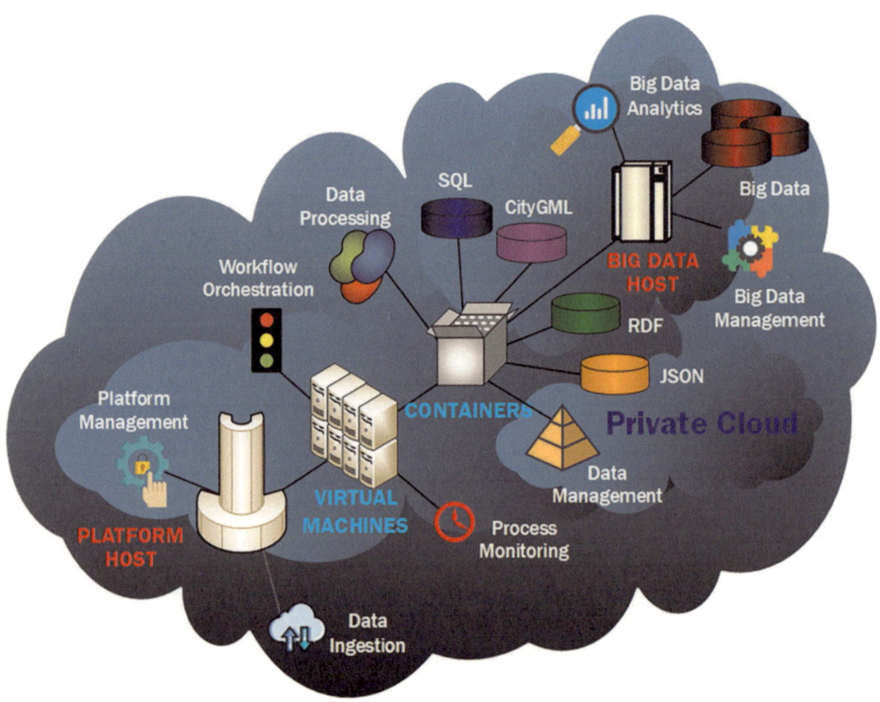

Figure 6.
System architecture of a cloud-based data platform (two hosts version).

- Data repositories (data specific but service independent) – database management systems for temporary data storage of data in four different formats – SQL (PostgreSQL), JSON (MongoDB), RDF (Neo4J) and CityGML (**3DCityDB**).

- Data management and data processing (data and service-specific) – bespoke software components developed to meet the specific requirements of the applications for managing and processing the data along its entire lifeline.

- Big data cluster – it incorporates the ecosystem for Big Data management and analysis. In our case, we used tools from Hadoop ecosystem, but they can be substituted or complemented with tools from other ecosystems. The use of Spark or Storm for analysis is not restrictive, either, since Hadoop can run directly Python and Java.

All software components are based on software without license costs from the public domain and community edition of enterprise products, selected to provide support for development, deployment and operation as discussed earlier. Their choices can be easily adapted to meet the specific requirements of the organizations.

Although the proposed architecture cannot match the universality, scalability and extendibility of public cloud provisioning, it works sufficiently well for many customers due to its simplicity and flexibility. The separation of platform management from data management, and the temporary data from the Big Data leads to an open architecture, highly scalable and extendable on both platform and layer levels. At the Cyber Security Research Centre of London Metropolitan University, the Big Data cluster is installed on a single host and accommodates only 80 TB of data, while at GATE Institute of Sofia University, it spans many physical hosts with a total memory capacity measured in petabytes. At the same time, at the Cyber Security Research Centre, all other components of the platform are installed on a single physical host, while at GATE Institute the separate groups of components are installed on different hosts to support working on multiple projects in parallel and to handle much larger amounts of data.

3.3 Setting up the deployment context of software components

Unlike ordinary information systems, which operate isolated from other software systems, the applications running on the platform consist of components which are dependent on each other within their context of execution. The data in different applications may come from the same data sources, while different data processing pipelines may use the services provided by the same components. This leads to the need for putting suitable control mechanisms in place for communication *tracking*, data *access control*, component *isolation*, process *synchronization and* workflow *orchestration*. The virtualization and containerization mechanisms on the cloud support a variety of options for achieving this. **Table 4** presents the alternative deployment contexts available for both the design of platform components used in all applications, as well as for the design of specific application components in different projects.

In SOA, each software component can be seen as a server, accessible through a dedicated TCP/IP port and exposing services on a different level – administration, implementation, and operation and at different stages of working – designing, developing, deploying and using. From this perspective, the TCP/IP ports of the engines executing the component services can be opened using three methods:

- Static IP address on the Internet, if needed, to be accessible by the physical users.

- Dynamic IP address, valid locally only and accessed by the different tools installed on the operating systems of the host computers or the VMs running on them.

- Dynamic IP address accessed programmatically from within the permitted VMs or containers of the platform according to the deployment of the software component.

The choice of a suitable context for component deployment can be made at design time, but since it may significantly affect the development, it must be made carefully. There are several design considerations to be accounted for, the most important being the degree of sharing and isolation of the services. Installing the components directly under the control of the operating system (*"bare metal"*) provides the widest possibility for sharing, but at the same time, it guarantees only very low isolation, while containerizing them provides the highest level of isolation but lowers the possibility for synchronization. An additional consideration which may need to be accounted for during the design is the visibility of the communications and the need to protect the information privacy and operation security. The closer to the operating system, the less control mechanisms can be used, while encapsulating the components within containers provides multiple mechanisms for controlling them.

Engine	Function	Context	Software support
Operating system (OS)	Resource allocation	Hardware	**Linux, Windows, MacOS**
Hypervisor (HV)	Resource virtualization	OS	Linux **KVM**, VMWare **vSphere**, MS **Hyper-V**, Oracle **VBox**
Virtual machine (VM)	Resource isolation	OS or HV	Ubuntu **Linux**, MS **Windows**, Oracle **Solaris**, Apple **MacOS**
Container manager (CM)	Component virtualization	OS or VM	K8 **Kubernetes**, Apache **Mesos**, HashiCorp **Nomad**
Container (CNT)	Component isolation	OS, VM or CM	**Docker**, **LXC**, **Windows Containers**, **Podman**
Server (SRV)	Service isolation	OS, VM or CNT	**NodeJS**, **JupytherHub**, **MongoDB**, **GlassFish**, **PostgreSQL**, **Neo4J**
Runtime (RT)	Service execution	OS, VM, CNT or SRV	Language-specific (programme interpreters, script shells, etc.)
Component	Service	OS, VM, RT CNT or SRV	Task-specific (data management, data analysis, visualization, etc.)

Table 4.
Context of deployment of software components on the cloud.

3.4 Enforcing security policies

The last step is to open the TCP ports for access to the services. Since each component exposes a service, the access to platform services from external clients or internal components is through a corresponding port. This can be done using two methods – control of the port visibility in accordance with the security policies (see **Table 4**) and allocating operational rights to external users according to their profiles. The first requires service registration after component deployment and must be updated for each new application, while the second requires user profiling.

4. Pilot projects

In this section, we will present three pilot projects, which have been completed using the platform by a joint team of London Metropolitan University and GATE Institute in three different projects – Computer network security analysis [22], Environment pollution monitoring in Sofia [23], and Impact of environment pollution on public health in London [24]. The successful completion of the projects proves the viability of the concept and its potential for use by public organizations, NGOs, SMEs and academic institutions dealing with Big Data.

4.1 Computer network traffic analysis (real-time pilot)

This project was the first pilot test of the data platform implemented after the concept described previously. It was dedicated to the detection of unsolicited behaviour due to unauthorized intrusion and/or the presence of malicious software. The goal was to analyze the network traffic and the event logs in real time to detect potentially missed interventions, as well as to perform a secondary analysis by examining the network traffic over a longer period. As a data source, we used a staged environment, running applications with malicious software attached to them. The generated data was captured by network analysers and transported for further analysis on the cloud over two protocols – TCP for the network packets and MQTT for the logs generated within a staged environment.

We performed two different analytics – real-time pre-processing and Pearson correlation of the streams and packet classification and intervention recognition using machine learning algorithms. The correlation analysis did not produce very exciting results due to the insufficient data generated by the simulator and the different speeds of the streams, but the subsequent forensic analysis using standard machine learning algorithms (logistic regression, neural networks and SVM) and deep learning (we used seven-layer CNN) was very interesting. It showed that although deep learning definitively produces the best results, it might not be the most suitable since some of the classical machine learning algorithms, which are much simpler, produce almost as good results. In our case, the SVG algorithm produced results close to the results obtained using CNN (see **Table 5**).

In this pilot, we used Docker for containerization of MongoDB as data storage and several data management components – for transportation (Mosquitto and Kafka),

Method	No. packets	RST (%)	ACK (%)	SYN (%)	Avg (%)
Regression	6268	85	81	31	66
SVM	6268	94	84	96	91
NN	6268	88	72	94	85
CNN	45,000	90	91	92	91

Table 5.
Precision of different methods for prediction of network packets (in %).

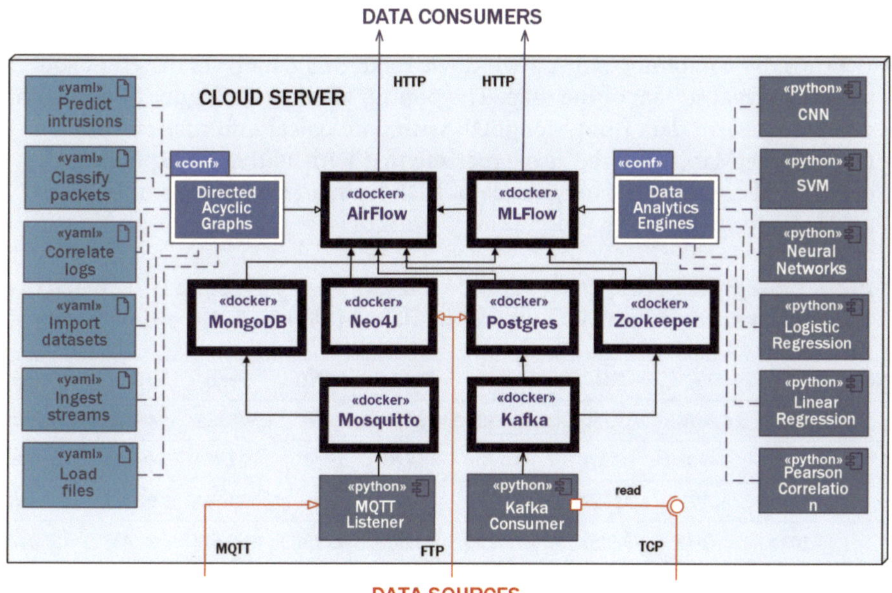

Figure 7.
Software components of the cloud-based data platform (portal host only).

operation synchronizing (Zookeeper), integration (Flask), workflow orchestration (AirFlow) and monitoring (MLFlow) – see **Figure 7**. Since we used only one physical host, the container management system did not play a significant role in the project.

The project was the first valuable test of the platform thanks to the data processing in both motion and peace, as well as the variety of methods for processing data both online and offline [22]. It is also interesting that due to the lockdown, the way in which the team of Cyber Security Research Centre of London Metropolitan University worked changed drastically – the staged environment was created earlier in London, while the actual platform for data processing is on the private cloud server of GATE Institute in Sofia, where the data was sent over the Internet. Both the development and the deployment of bespoke software were done remotely from London, which proves additionally the viability of the whole concept.

4.2 Environment pollution monitoring (Sofia pilot)

This project was focused on the environmental monitoring in Sofia due to the significant air pollution in the city [23]. The air pollution data was collected from 13 sensor stations across the city. The readings were formatted as JSON objects and sent to the cloud over MQTT for analysis. Before the data was accumulated in the MongoDB database, it was correlated using the standard Pearson algorithm for establishing dependence between different factors of pollution like temperature, humidity, gases, and particles in the air (see **Table 6**). The archived data from previous periods was uploaded to the same database using NiFi to train the machine learning algorithms before prediction.

An essential addition to the platform in this project was the integration of the Cesium GS Cesium JS component, which allows multi-layered visualization on top of 2D maps [25]. As a 2D base, we used the free service provided by the OpenStreetMap Foundation [26], which has worldwide coverage. We integrated and visualized a variety of data from different data sources using two methods:

- Pop-ups: By combining pop-up windows rendering data from different sources with the visual stream of the map. The pop-up window on **Figure 8**, for example, combines sensor data from MongoDB with ontological information from OpenStgreetMap. We have also experimented with adding an explicit ontological model of the urban area in RDF format extracted from the Neo4J graph database [24].

- Projections: By superimposing additional layers on top of the base 2D maps to add urban infrastructure and 3D building models. In **Figure 9**, for example, the 3D

No	CO	SO_2	NO_2	O_3	PM2	PM10	Press	Hum	Temp
1	0.5039	6.3676	15.730	43.096	5.8484	8.291	955.855	46.322	16.202
2	0.4969	5.4908	38.971	53.154	8.6820	15.797	933.432	46.311	16.578
3	0.5010	4.7727	11.779	47.642	3.5257	5.965	959.594	96.717	20.416
4	1.1032	8.7043	9.9153	37.223	7.3437	15.251	960.677	31.885	18.247
5	0.3916	6.8086	14.5090	35.469	6.9110	15.995	959.174	54.637	12.476

Table 6.
Positive correlation between outdoor temperature and humidity (Pearson 0.972).

DOI: http://dx.doi.org/10.5772/intechopen.1003268

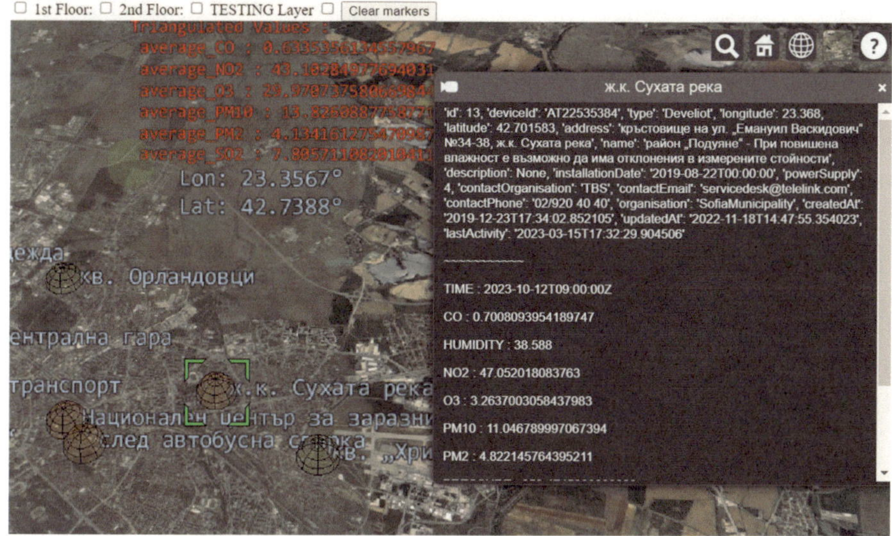

Figure 8.
Data integration and visualization of the air pollution in Sofia.

Figure 9.
Integrating 3D buildings model and ontological information with 2D map.

model of the building which has been reconstructed offline from its 2D floor plan, is subsequently embedded in the map to show the location of indoor sensors.

As a result of this integration, we can display a rich combination of data, coming from different sources in different formats [23]. This application has been live for more than a year and can be accessed over the Web at: http://194.141.1.61/.

In this pilot project, we also added several components for platform management, implemented using public domain and community edition software:

- Remote access: Proxmox
- Identity management and access control: NGinx
- Indexing and searching: Elasticsearch
- Auditing and reporting of data services and operations: Kibana

The main data sources of these components are the system logs generated on different levels of operation of the platform software – OS, VMs, containers, servers, and runtime engines (see **Table 4**). These additional components are application independent; their primary role is to enhance the control of the operations. This way, they prepare the migration of the platform pilots to commercial provision, as well as the use of the platform as data and service provider in future dataspaces [4].

4.3 Impact of environment pollution on public health (London pilot)

This project started as a mirror of the project for monitoring the air pollution in Sofia. We first implemented most of the functionality we had in Sofia, this time on the private cloud of the Cyber Security Research Centre of London Metropolitan University (see **Figure 10**). The real-time sensor data we were ingesting from the stations in London came from 130 locations, so the amount of data was around 10 times bigger than in Sofia, although, in terms of real-time, it is still manageable. Because of this, the sensor data was initially gathered in the MongoDB database in JSON format for preliminary analysis. Further, we were able to collect offline archived data from the last 5 years in a structured format (CSV files), which was transferred to the server and stored initially in PostgreSQL database for trends analysis. Finally, we implemented our own sensor station to collect information about indoor pollution and used its output to correlate the indoor and outdoor factors in the area around the building of the university [26]. The correlation

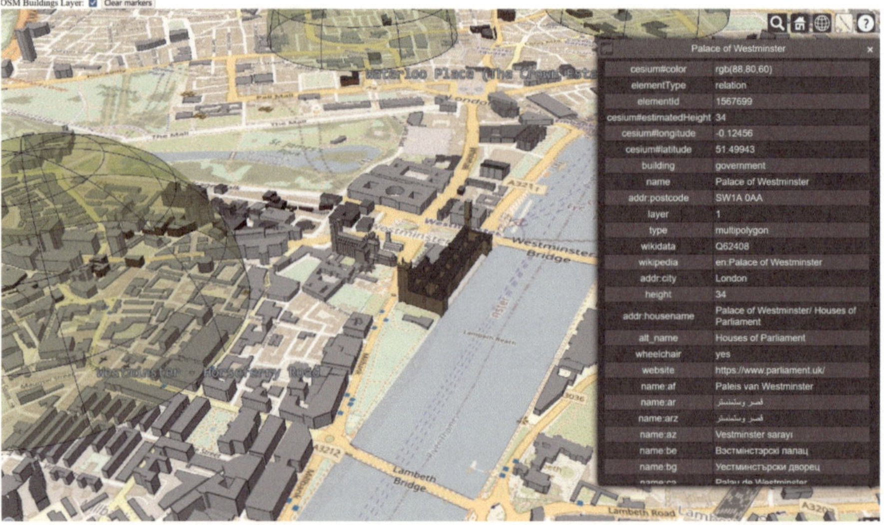

Figure 10.
Data integration and visualization of the air pollution in London.

showed a very close dependence between indoor pollution and outdoor pollution with some rare outburst which activities inside the building can explain.

Further expansion of the platform was achieved thanks to the addition of a new data source with data about medication prescriptions, freely available from one of the sites of UK National Health Service (NHS). Due to the volume of this data, we transferred the entire dataset we obtained from NHS using NiFi directly to the Hadoop cluster. To enable further use and more complex analysis of all data periodically all the temporary data stored in MongoDB was transferred to the Hadoop cluster using NiFi. The two datasets – the air pollution readings and the prescription data – were then joined in Hadoop and analyzed using Spark for possible correlation between the pollution and the respiratory diseases, which were categorized on the base of the prescribed medications. The analysis was performed using three different methods of calculating the correlation – Pearson, Spearman and Kendall, with comparable results. **Figure 11** presents some of the results of this analysis for a fragment of the dataset of prescriptions against the category of respiratory diseases. As it is clearly visible from the diagrams, there is a strong correlation between most pollutants and respiratory diseases, with NO_2 predictably being the most harmful. This pilot has been live for several months and is available on the Web at: http://217.38.61.107:5000/.

The most significant addition to the platform in this project was the full utilization of Big Data technologies for cross-domain data analysis. The Hadoop cluster for managing the Big Data was created on three virtual machines, emulating hosts for one name node and two data nodes accommodating the data in Hadoop HDF memory space. The Big Data analytics was performed in a distributed environment with a Spark server operating on the name node and two Spark clients operating on the data nodes. The project is currently still under way and will continue analyzing the data for revealing further dependencies as well as the dynamic of changes depending on the environment conditions and the period.

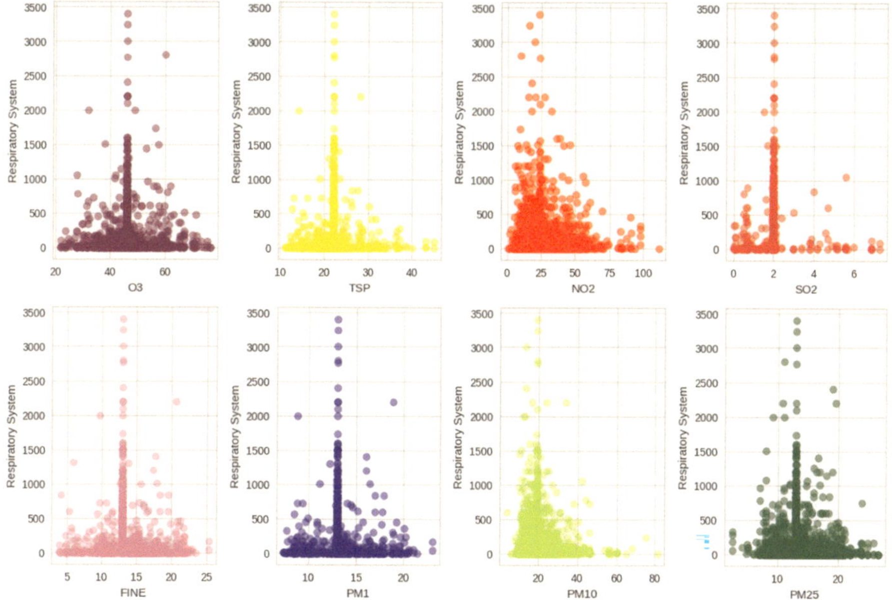

Figure 11.
Correlation between air pollutants and respiratory diseases in greater London.

Virtualization	Containerization	Orchestration
• *Heterogeneity* of hardware and system software	• *Modularization* of the software with no dependencies to set	• Support for *reusability* of existing solutions in process workflows
• *Scalability* of devices, memory and users	• *Efficiency* in memory, CPU and storage usage	• *Model-driven* application development
• *Choice* of convenient computational environment	• *Portability* of containers across platforms without code changes	• Support for *auditing* of monitoring, analyzing, and billing purposes
• *Transparency* of the physical location of the data and services	• Supporting *configuration generation* using templates	• *Reproducibility* of operations by preserving dependencies
• *Centralization* of the system administration and maintenance	• Full *traceability* of the operations for testing and debugging purposes	• Possibility for *process automation* based on planning heuristics

Table 7.
Advantages of cloud-based data platform for Big Data processing.

5. The gains and the burdens of being independent

Big Data platform on the cloud has many advantages for public and private businesses, large and small organizations, academia and industry (see **Table 7**).

Operationally, platforms on the premises provide more limited opportunities in comparison with the platforms of the public clouds. For example, most public clouds support the training of machine learning algorithms, which drastically reduces the *time* and *computational resources* needed for training. However, the private clouds retain the *data ownership* and protect the *privacy* of data and operations.

The software without license fees brings some disadvantages, too. Custom-built platforms may incur higher *maintenance costs* due to the need to hire highly qualified staff or even external consultants. But this drawback diminishes quickly with multiple projects because of the *cost spreading*, while the running costs on the public clouds are cumulative and, as a result, are higher per project. From a strategic perspective, in-house operation has an additional advantage – it stimulates the local economy by fostering *independence* from the commercial software vendors and service providers. In balance, this solution is definitively a better choice for project oriented organizations and software houses with extensive in-house development.

6. Possible extensions and directions for future work

The concept of a data platform for Big Data processing on a private cloud using free software proved itself strongly through the pilots. Our work in this direction can continue to make the platform more valuable for both the organizations which employ it as a development environment and for clients which rely on its services:

- Methodology adoption: The DevOps employs shared repositories for various purposes. Previously, we have used GitLab for automatic creation of computing infrastructure, component deployment and integration on the public cloud [20]. It significantly increases the productivity and improves the quality. It can be employed on the private cloud as well. This is valuable for software vendors and consultancy organizations in which multiple teams are working in parallel, sharing both data and software.

- Analysis hybridization: The open architecture allows the addition of more components, particularly for hybridization of the analytics. We have already done some work on hybrid analytics by combining pure data analysis with knowledge-based reasoning [16]. This would allow cross-domain analysis beyond the simple data analysis since it can utilize the dependencies within the data formulated in heuristic rules.

- Workflow automation: We have also conducted some experiments for the generation of Docker and Airflow, configuration files, based on the ontological model of the platform (see **Figure 3**). This can be done using templates, which can also be used for explanation generation, accounting for the causal links between tasks and results [21].

- Service homogenization: Combining bespoke and universal components may lead to granularity problems. For example, we were trying to implement the access control using Fiware Keyrock [27], but it required too many additional components, so we used Nginx instead. On the other hand, for the enrichment of data in tabular format, we wanted to use the **Grafterizer** [28], but for non-relational data, we needed other components, so we opted out in favour of a more universal bespoke development.

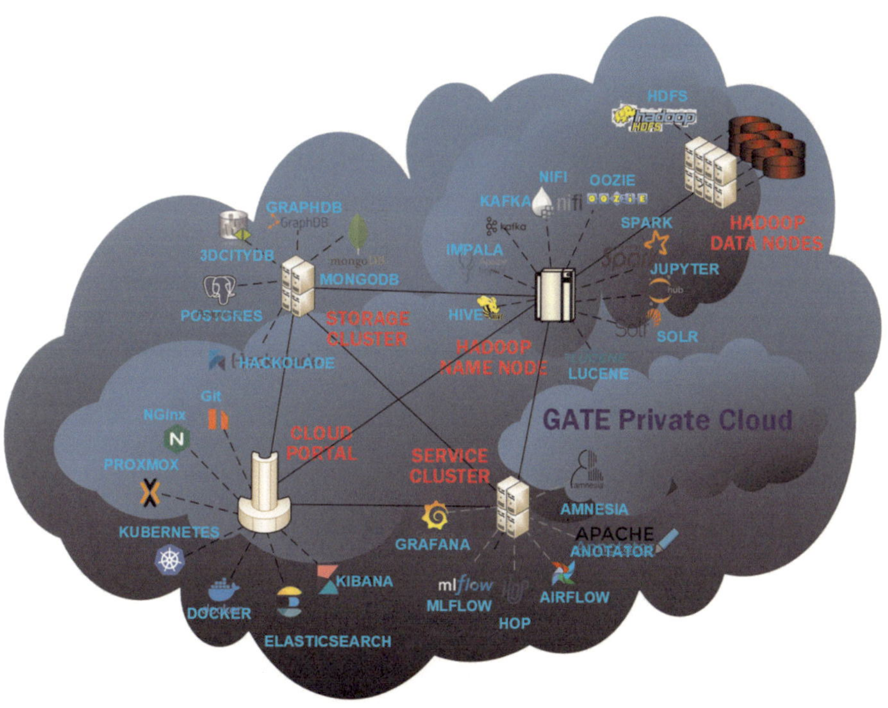

Figure 12.
Architecture of GATE big data platform.

With the growth of digital data, the need for utilization of data platforms becomes more important by the day. The open and highly scalable architecture of a platform built on private clouds using software without a license fee can easily evolve into an enterprise solution, running on multiple physical hosts and managing petabytes of data. This guarantees scalability without the need for significant initial investment. For example, **Figure 12** shows the enterprise version of the GATE Data Platform, built this way—not only a viable but also a highly lucrative option.

Acknowledgements

The work reported here has been conducted over a period of more than 5 years in collaboration with colleagues, students and industrial partners from London Metropolitan University in the United Kingdom and Sofia University in Bulgaria. It has been made possible thanks to several grants from the United Kingdom (Innovate UK and HEIF), the Bulgarian government (OPNOIR Grant Agreement No. BG05M2OP001-1.003-0002-C01) and the EU H2020 Research Framework (WIDE-SPREAD 2018–2020 Grant Agreement No. 857155). It has also been supported by private companies in the United Kingdom (Lloyds Banking Group, Oxagon) and Bulgaria (Rila Solutions, Ontotext), providing additional resources for platform development. The authors are greatly thankful to all collaborators and partners for their professional help, continuing support and unwavering trust over the years. All recommendations and considerations, however, are purely of the authors and should not be considered the official policy of these organizations.

Author details

Vassil Vassilev[1,2*], Viktor Sowinski-Mydlarz[1,2], Pawel Gasiorowski[1,2], Sorin Radu[2], Sabin Nakarmi[1,2], Martin Hristev[1], Reza Baghaeishiva[1] and Tarun Bali[2]

1 Cyber Security Research Centre – London Metropolitan University, UK

2 GATE Institute – Sofia University, Bulgaria

*Address all correspondence to: v.vassilev@londonmet.ac.uk

References

[1] Gartner, Inc. 10 top strategic technology trends [Internet]. 2023. Available from: https://www.gartner.com/en/information-technology/ [Accessed: July 06, 2023]

[2] Moses B, Gavish L. What is a data platform? [Internet]. 2023. Available from: https://www.montecarlodata.com/ [Accessed: July 07, 2023]

[3] Strong A. Containerization vs. virtualization: What is the difference? [Internet]. 2022. Available from: https://www.burwood.com/blog-archive/ [Accessed: July 07, 2023]

[4] Anjomshoaa A et al. Data platforms for data spaces. In: Curry E et al., editors. Data Spaces. Cham: Springer; 2022. DOI: 10.1007/978-3-030-98636-0_3

[5] IBM. IBM storage scale Big Data and analytics support [Internet]. 2023. Available from: https://www.ibm.com/docs/en/storage-scale-bda [Accessed: July 07, 2023]

[6] Hewlett-Packard Enterprise. HPE Ezmeral Data Fabric [Internet]. 2023. Available from: https://www.hpe.com/us/en/hpe-ezmeral-data-fabric.html [Accessed: July 07, 2023]

[7] Oracle. Oracle Big Data Appliance [Internet]. 2023. Available from: https://docs.oracle.com/en/bigdata/big-data-appliance/index.html [Accessed: July 07, 2023]

[8] Amazon Web Services, Inc. Amazon EMR [Internet]. 2023. Available from: https://aws.amazon.com/emr/ [Accessed: July 07, 2023]

[9] SAP. SAP HANA Cloud [Internet]. 2023. Available from: https://www.sap.com/uk/ products/technology-platform/hana.html [Accessed: July 07, 2023]

[10] Cloudera, Inc. Cloudera Data Platform [Internet]. 2023. Available from: https://www.cloudera.com/products/cloudera-data-platform.html [Accessed: July 07, 2023]

[11] Kunigk J, Buss I, Wilkinson P, George L. Architecting Modern Data Platforms. 1st ed. Sebastopol: O'Reilly; 2019. p. 640

[12] Amazon Web Services, Inc. AWS Lake Formation [Internet. 2022. Available from: https://aws.amazon.com/lake-formation/?c=a&sec=uc3 [Accessed: July 07, 2023]

[13] Google. Cloud data warehouse to power your data-driven innovation [Internet]. 2023. Available from: https://cloud.google.com/bigquery/ [Accessed: July 07, 2023]

[14] Microsoft. Azure Databricks [Internet]. 2023. Available from: https://azure.microsoft.com/en-gb/products/databricks [Accessed: July 07, 2023]

[15] Almeida A, Brás S, Sargento S, Pinto FC. Time series big data: A survey on data stream frameworks, analysis and algorithms. Journal of Big Data. 2023; **10**(1):83. DOI: 10.1186/s40537-023-00760-1

[16] White T. Hadoop. 4th ed. Sebastopol: O'Reilly; 2015. p. 754

[17] Taylor D. Top 15 Big Data tools and software [Internet]. 2023. Available from: https://www.guru99.com/big-data-tools.html [Accessed: November 07, 2023]

[18] Chambers B, Zaharia M. The Definitive Guide. 1st ed. Sebastopol: O'Reilly; 2018. p. 603

[19] Vassilev V, Sowinski-Mydlarz V, et al. Intelligence graphs for threat intelligence and security policy validation. In: Bansal P et al., editors. Intelligent Systems and Computing. Vol. 1164. Springer; 2020. pp. 125-139. DOI: 10.1007/978-981-15-4992-2_13

[20] Vassilev V, Phipps A, Lane M, et al. Two-factor authentication for voice assistance in digital banking using public cloud services. In: Proc. 10th Int. Conf. Confluence. Noida, India: IEEE; 2020. pp. 404-409. DOI: 10.1109/Confluence47617.2020.9058332

[21] Vassilev V, Ilieva S, Sowinski-Mydlarz V, et al. AI-based hybrid data platforms. In: Curry E et al., editors. Data Spaces. Springer; 2022. pp. 147-170

[22] Vassilev V, Ouazzane K, Sowinski-Mydlarz V, et al. Network security analytics on the cloud: Public vs. private case. In: Proc. 13th Int. Conf. Confluence. Noida, India: IEEE; 2023. pp. 151-156. DOI: 10.1109/Confluence56041.2023.10048889

[23] Vassilev V, Sowinski-Mydlarz V, Mariyanayagam D, et al. Towards first urban data space in Bulgaria. In: Proc. IEEE Int. Smart Cities Conference. Paphos, Cyprus: IEEE; 2022. pp. 1-7. DOI: 10.1109/ISC255366.2022.9922237

[24] Vassilev V, Virdee B, Ouazzane K, et al. Data platform and urban data services on private cloud. In: Zghang Y et al., editors. Smart Trends in Computing and Communications. Vol. 650. Springer LNNS; 2023. pp. 263-275. DOI: 10.1007/978-981-99-0838-7_23

[25] Cesium GS, Inc. The platform for 3D geospatial [Internet]. 2023. Available from: https://cesium.com/ [Accessed: November 07, 2023]

[26] OpenStreetMap Foundation. Planet OSM [Internet]. 2023. Available from: https://planet.openstreetmap.org/data [Accessed: November 07, 2023]

[27] Fiware. Keyrock Identity Manager [Internet]. 2023. Available from: https://keyrockfiware.github.io/ [Accessed: July 14, 2023]

[28] Stiftelsen S. Grafterizer 2.0 [Internet]. 2023. Available from: https://www.eubusinessgraph.eu/grafterizer-2-0/ [Accessed: November 07, 2023]

Chapter 5

Perspective Chapter: Open Science Rejuvenation with AI – The Past, Present and Future Dimensions

Mayukh Sarkar and Sruti Biswas

Abstract

The inception of Open Science ideology originated with a vision towards advancing the scientific knowledge with the value of availability, accessibility, reusability, and transparency to democratise complete research cycle across all sectors of society irrespective of any class or community has successively coalesced with various vistas of "Open movement" and also outreached its realm from STEM subjects to the universe of disciplines. The advent of Artificial Intelligence (AI) with machine learning (ML) and its specific specialisations like deep learning (DL), reinforcement learning (RL) and genetic algorithms (GA) enunciate an intelligent, expert, and decision support system revolutionises the contemporary technologies to a newfangled one, providing the most powerful discovery engine for analysis, retrieval, transfer of data, hypothesis/metrics generation, and determining research originality open up new opportunities in the domain of Open Science as well as eroding the commercial interests of the enterprises. The chapter, therefore, portrays the symbiosis of Open Science and AI in the canvases of historical antecedents how it evolving progressively, instigates the AI drivers (ML, DL, RL, and GA) and enablers (natural language processing, computer vision, ontology and knowledge graph) practicable in Open Science, evaluate recent Open Science and AI amends of global confederations.

Keywords: Open Science, artificial intelligence (AI), machine learning (ML), deep learning (DL), reinforcement learning (RL), genetic algorithms (GA)

1. Introduction

Through the microscopic evaluation of various encyclopaedic, dictionary and scholastic definitions of the term "Science" derived from the Latin word "Scientia", meaning "knowledge", refers to the human holistic endeavour to understand the natural and other unexplainable phenomenons of the universe with unique logical method containing hypothesis construction, experimentation, observation, analysis and deriving results, i.e. diverse and varied from one another evolved with the time in order to establish the theory. As Segan [1] rightly mentioned, "A central idea of science is that to understand complex issues (or even simple ones), we must try to free our minds of dogma and to guarantee the freedom to publish, to contradict and to experiment. Arguments from authority are unacceptable." In the modern dilemma of data and the information-oriented world, the fascinating notion of "Openness", where "Open means anyone can freely access, use, modify, and share for any purpose (subject, at most, to requirements that preserve provenance and openness)" as concretely defined by Open Knowledge Foundation [2] from the perspective of data and content. Nevertheless, in reality, it is a multifaceted jargon associated with the global "Open movement" beyond any context-based definitions. Now merging these two distinct concepts, Open Science emerged as a new field of study and became an inextricable component of the "Open movement", which began when scientific journals were published in the early seventeenth century [3]. The Facilitate Open Science Training for European Research project has defined "Open Science", as "the practice of science in such a way that others can collaborate and contribute, where research data, lab notes and other research processes are freely available, under terms that enable reuse, redistribution and reproduction of the research and its underlying data and methods" [4]. According to European Commission [5], "Open Science is a system change allowing for better science through open and collaborative ways of producing and sharing knowledge and data, as early as possible in the research process, and for communicating and sharing results." According to UNESCO's [6] draft recommendation, "Open Science an inclusive construct that combines various movements and practices aiming to make multilingual scientific knowledge openly available, accessible and reusable for everyone, to increase scientific collaborations and sharing of information for the benefits of science and society, and to open the processes of scientific knowledge creation, evaluation and communication to societal actors beyond the traditional scientific community."

Though the inception of the Open Science movement began with the goal of promoting scientific knowledge in fields like modern STEM (science, technology, engineering, and math) subjects, its true essence has now expanded to encompass a wide range of fields, all under the umbrella term "Open Science." So, comprehending the standard definitions, Open Science refers to the noble and evolving ideas with a vision towards advancing the scientific knowledge from the realm of STEM subjects to the universe of disciplines with the value of availability, accessibility, reusability, and transparency to democratise complete research cycle across all sectors of society beyond borders, paywalls, intellectual patterns irrespective of any class or community successively coalesced with various vistas of "*Open movement*" that include Open Access, Open Source, Open Data/FAIR Data/Open Knowledge, Open Research/Methodology, Open Scholarships, Open Peer Review (also includes Open Identities and Open Interactions), Open Metrics/Impact, Open policies, and Open Educational Resources.

In order to understand Artificial Intelligence (AI), first, let us understand the concept of intelligence (preferably human or other sentient creatures), which refers to the "mental quality that consists of the abilities to learn from experience, adapt to new situations, understand and handle abstract concepts, and use knowledge to manipulate one's environment [7]." According to Poole and Mackworth [8], AI is the specific domain of study syntheses and analyses the "*computational agents*" (whose activities are described in terms of human or computer-based computations) that "*act intelligently*" (gets smarter as it goes along, acquiring both the short-term and long-term effects of its activities while making decisions, adapting to different situations and circumstances, and selecting the most appropriate options). So, AI is the evolving agent system simulating human intelligence in machines partially, entirely or goes beyond the intellectual capability of the human brain to acquire and apply all the cognitive functions as referred to in the opening statements of this paragraph, unleash new possibilities in various domains.

The advent of AI with machine learning (ML) and its specific specialisations like deep learning (DL), reinforcement learning (RL) and genetic algorithms (GA) enunciate an intelligent, expert, and decision support system that revolutionises the contemporary technologies to a newfangled one, providing the most powerful discovery engine for analysis, retrieval, transfer of data, hypothesis/metrics generation, and determining research originality opens up new opportunities in the domain of Open Science as well as transforming the commercial interests of the enterprises. The rapid development in AI research and its rising applications in Open Science urges a need for reformation in policy frameworks congruence with global AI policies. The chapter, therefore, portrays the symbiosis of Open Science and AI in the canvases of historical antecedents and how it is evolving progressively, instigates the AI drivers (ML, DL, RL, and GA) practicable in Open Science, and evaluates recent Open Science and AI amends of global confederations (European Commission, UNESCO, COS, OECD) would help policymakers to recommend a holistic policy guideline for amalgamating two notions detouring the ethical issues related to them.

2. Open Science with AI: the beginning

Identifying the origin of AI's Open Science application is challenging to decipher. Though Open Science and AI developed over a lengthy period, they only came together in the last two or three decades, as reflected through the "Guerilla Open Access Manifesto" [9]. Following that, in the same year, the concept of digital shadow libraries was born and widely practised all over the globe in the form of "Library Genesis" or "LibGen", an initiative of Russian scientists went online [10]. In 2011, Kazakhstani programmer Alexandra Elbakyan, Aron Swartz's condign successor, pushed the Open Access campaign to the next level by breaching the paywall shield, extending his alterations to the meaning of accessing academic literature, and creating Sci-Hub [11] (a PHP-coded online application) soon introduced new databases, servers and multiple mirrors able to interlink the databases of LibGen. Considering several obstacles such as legal actions, hackers' attacks, frequently blocked by ISPs/ Operators, and changing domains, both the applications are still functioning, e.g., Sci-Hub's logs analysis displayed 28 million download requests from September 1, 2015, to February 29, 2016 [12], and more than 1.2 million new records were added to the LibGen database between January 2008 and April 2014 [13]. According to a research analysis of 2750 random samples from 55 databases by Houle [14], the full-text retrieval rate for Sci-Hub and LibGen reported 70% and 69%, respectively, for April 2017. This trend continues into the year 2023, with academics all over the world using Sci-Hub; among 6632 medical students from six Latin American countries (Argentina, Bolivia, Chile, Colombia, Paraguay, and Peru), 10.3% used Sci-Hub at least once a week to consult scientific journals, making it a particularly notable example [15], and the same backed by Ajani et al. [16] by addressing it as "a blessing in disguise to library users." In addition, according to data generated by Sci-Hub [17] over the past month, the top five nations in terms of downloads are China (45.65 M), the United States (27.28 M), Brazil (7.67 M), India (2.79 M), and Russia (2.79 M).

Besides the guerilla movement for OA to information (recall Aron Swartzs' [9] famous quote "information is power"), another silent revolution happened to emphasise data (as mathematician Clive Humby [18] used the phrase "data is the new oil") in the form of ArnetMiner, an intelligent system for the academic community and an AI application to Open Science that intends to extract and mine scholarly social networks in five major steps; (i) auto-extraction of researchers' profiles and their publication data through web-mining, (ii) integrating them using name identifier, (iii) storing and indexing the data using MySQL and inverted file index, respectively, (iv) data modelling, and (v) executing search results [19].

3. Open Science with AI: the present

The present paradigm of Open Science revolves around different AI-based applications that align with Open Science goals (publishing and disseminating research ideas efficiently, removing linguistic and sharing hindrances, encouraging transparency/originality in research and supporting the collaborative networks), and assisting global academicians through developing various intelligent tools. Now each of these goals can be satisfied through AI, for instance, auto-summarisation, identifying and predicting emerging areas of interest for research (data science and data analytics), generating new book ideas (natural language generation), estimating book performance (predicting the potential essence of the book before it is written), checking language/grammar, editing manuscripts, extracting keyword to sentence/sentence to keywords, managing references, making the scholastic community comfortable to publish and communicate research. Machine-generated books/literature, auto-translation, workflow suggestions (recommendation systems), text-2-speech and video transcriptions (NLG), and sharing the best practices attempt to remove the linguistic and sharing barriers. Checking plagiarism, research integrity, bias recognition, natural language processing, and sentiment analysis promote transparency. Finally, AI supports the collaborative networks by identifying potential authors (BAIT), peer reviewers, editors and editorial board members, matching topics and stakeholders using recommendation systems, and identifying specific communities or researcher groups online using sentiment analysis.

In tandem with the discussed context, preprints and author services driven by the futuristic vision of Open Science are becoming more prominent on enterprise/-community-oriented platforms, which have significantly increased their visibility. Research Square (https://www.researchsquare.com/) and its subdivision American Journal Experts (https://www.aje.com/#), is one of the pioneer enterprises that acquired expertise in providing the services described earlier without violating the ethical grounds. Furthermore, IntechOpen (https://www.intechopen.com/) is an example of a significant industrial initiative in support of Open Science, a book and journal publishing house that believes and adheres to the OA/Open Science principles (recognised by the Budapest Initiative, International Association of STM Publishers; ALPSP; COPE; CC; Crossref; OASPA) and is indexed in the prominent platforms (WoS, Scopus, BKCI, BIOSIS Previews, Zoological Record etc.). Another OA journal publishing platform Frontiers Media (https://www.frontiersin.org/) also leveraged Open Science with AI, using a state-of-the-art platform for peer review, semantic algorithms, an extensive reviewer database with an article-level metrics and a researcher profile for high visibility, as well as a streamlined production workflow and a digital editorial office. Because of this trend towards greater openness and access to research ideas among researchers, modern enterprises have reformed their business interests to keep up with the changing surroundings. For example, most leading publishing houses, such as Elsevier, Emerald, Sage, Springer, Taylor & Francis, Willy etc., started providing an OA publishing option to the academicians, which is also considered a dilemma of change towards welcoming the open science movement.

eLife Science Publications Ltd. (https://elifesciences.org/), an online non-profit OA journal publishing house for life sciences and biomedical research, developed a platform for research communication that attempts to speed up the discovery process by using AI. Since 2017 they have designed two noteworthy projects, ScienceBeam [20, 21] and PeerScout [22]. With Apache Beam and TensorFlow, ScienceBeam seeks to unlock the PDF format's immense store of scientific knowledge and generate a complete XML document using computer vision and natural language processing. PeerScout uses machine learning trained API and NLP to locate relevant peer reviewers from an existing collection based on the qualities of the articles they have reviewed and authored. Besides these two landmarks, NLP and general ML approaches have been used in other projects to analyse citations' context and sentiment. Their technology and data science team construct their products in an open-source environment and make them available on GitHub under permissive open-source licences [23].

Wang [24] has observed a groundbreaking investigation where the Microsoft research group aimed to scour the Web for research artefacts and pulls the concurrent scholarly information represented through a web-scale heterogeneous entity graph/ knowledge graph (known as Microsoft Academic graph or MAG) using AI-powered agents trained in natural language understanding and reinforcement learning [25, 26]. The research empowered Microsoft to launch an analytic, discovery and auto-distribution service called Microsoft Academic Services (MAS), integrating MAG, Azure Storage account, Microsoft Academic Knowledge API and Microsoft Academic Knowledge Exploration Service (MAKES). An improvised research version was released in 2017 under the ODC-BY licence, which interlinked both the MAG and ArnetMiner, known as the Open Academic Graph (OAG), a large-scale linked graph (dataset of 0.7 billion entities and 2 billion relationships); looking forward to investigating citation networks, collaboration, content analysis, and finding the answer to how diverse academic graphs interact with one another. The developers presented a unified LinKG framework with three linking modules (a. venue name matching and sequence encoding, b. hashing technique and convolution neural networks, and c. heterogeneous graph attention network technique) for three entities (a. venue, b. paper and c. author) that can efficiently handle and bypass the challenge of designing a large-scale linked entity graph [27]. The project is still under constant development (OAG v.2.1 released in November 2020), and a new affiliation entity has been incorporated along with the previous entities and generated 16,384, 29,948, 119,384,813, and 1,829,385 linking relations among the two graphs for affiliation, venue, paper and author entity, respectively.

The subsections integrating Open Science with AI under two distinct terms, "drivers" and "enablers", represent different aspects influencing AI technologies' development, adoption, and growth. AI drivers are the major factors that trigger AI technologies' advancement, application, and integration. On the other hand, AI enablers are the factors, essentials, or resources that create an environment conducive to the development, utilisation and support of the successful implementation and growth of AI technologies.

3.1 AI drivers for Open Science

Since the dawn of the new millennium, breakthroughs in AI research fields have seen multiple rounds of rapid advancement; even each of the noteworthy field's subfields has become a specialised area. From the foundation of AI in the Dartmouth College workshop in 1956, the field got richer in terms of establishing theory and various learning algorithms [28]. It is not an easy undertaking to foster each AI breakthrough from the very foundation, while expansion of Open Science is also in progress. Keeping this inadmissible issue in mind, we have only instigated the AI drivers practicable in Open Science. Here we have mentioned a few that have been used or are still in the experimental pipeline to contribute to Open Science and achieve beyond its goals.

3.1.1 Machine learning (ML)

Machine learning (ML) refers to such computational algorithms, approaches to identify the hypothesis from the vast and complex space of possibilities losing minor data points by simulating human/sentient creature's intelligence and adopting the environment [29]. This task might carry forward by recognising the patterns in the input data/entities through several learning mechanisms; *supervised learning* (each training sample/characteristic of input data is linked with its known categorisation label), *unsupervised learning* (based on the input data used for training, the algorithm determines its own path and moves towards perfections with each attempt) and *semi-supervised learning* (uses both labelled and unlabeled data, where labelled section can help learn the unmarked part). Further, the ML algorithms are divided into distinct classes depending on the type of input, learning process and genre of the model [30].

3.1.2 Deep learning (DL)

Deep learning (DL) is a subset of ML that relies on artificial neural networks (ANN) to learn several representations simultaneously. DL systems refer to "a class of multi-layered networks capable of automatically learning meaningful hierarchical representations from various structured and unstructured data" on the cutting edge of ML innovation [31]. DL advances have made it possible to construct novel representations, information extraction, and inference speculation from complex data sources like photos, videos, texts, speeches, time series, and other discrete events. Different ANN algorithms such as perception neural network (PNN), back propagation (BP), self-organising network (SON), self-organising map (SOM), and learning vector quantisation (LVQ) work at the backend of a DL structure with its common algorithms namely, restricted Boltzman machine (RBN), deep belief network (DBN), convolutional neural network (CNN), and stacked auto-encoder (S-AE) etc. [32].

3.1.3 Reinforcement learning (RL)

Reinforcement learning (RL), as a specific flavour of ML, deals with an AI-powered agent system's state at distinct time points and examines situations in which the assumptions made by a hypothesis influence the formation of new data points without accessing the labelled ones. It is different from traditional ML techniques (which use supervised and unsupervised learning mechanisms) and must be able to sense the environment and take action against the entire problem importing equal importance to a goal-directed objective and an uncertain environment [33]. Tabular TD(0), tabular TD(λ), TD(λ) with linear function approximation, every-visit Monte-Carlo, Gradient temporal difference (GTD2), Least-squares temporal difference

(LSTD), Least-squares policy evaluation (LSPE), PAC-MDP, SARSA(λ) with linear function approximation are some effective RL algorithms [34].

3.1.4 Genetic algorithms (GA)

Using evolutionary modelling, genetic algorithms (GA) create and alter (artificial evolution) a search algorithm approaching both heuristic and metaheuristic search based on a set of natural selection and genetic mechanisms involving the cycle of initialisation, crossover, mutation, fitness computation, selection, and termination to execute Darwin's survival of the fittest principles [35, 36]. Basically, GA is an invariant of "*evolutionary algorithms*" (performing complex optimisation or neural network learning process and replacing the solution with the preferable one with the flexibility to evolve in different situations) and "*evolutionary intelligence*" (ability to overextend and integrate present intelligence to allow for new ones), a part of the broader field of "*evolutionary computations*", focuses on resolving the encoding issues, handling constraints and introducing DNA computing [37–39].

3.2 AI enablers for Open Science

In addition to above mentioned AI drivers, some enablers can leverage Open Science as future-proof. We will go over a few of the key enablers now:

3.2.1 Natural language processing (NLP)

In computer science and human linguistics, natural language processing (NLP) refers to the machines' ability to recognise and understand human text language with multi-directional applications such as machine translation, speech recognition, grammatical and semantic analysis, sentiment analysis, text summarisation, information extraction, text and audio generation, text-to-speech and speech-to-text conversion, question-answering system, dialogue system, chatbots, and voicebots [40]. In mid-2017, researchers at Google introduced "Transformers", a novel architecture combining three potential mechanisms called an encoder-decoder framework, attention, and transfer learning became benchmark research in NLP. The two popular transformers, Generative Pretrained Transformer (GPT) and Bidirectional Encoder Representations from Transformers (BERT), with their latest upgrades like GPT-Neo, and GPT-J or DeBERTa, are now in the mainstream [41].

3.2.2 Computer vision (CV)

The domain of computer vision (CV) deals with studying and developing specific algorithms that enable computers to identify, process, analyse, and interpret digital objects with visual properties such as images/videos. As rightly mentioned by Bekhit [42], "if AI enables computers to think, computer vision enables them to see, observe, and understand." The entire methodology is carried out in multifarious stages using several techniques, scilicet, (i) image preprocessing, which includes grayscale manipulation, edge enhancement and detection, noise removal, image restoration, interpolation, (ii) image segmentation, (iii) image processing includes feature extraction, texture analysis, pattern recognition, and (iv) image classification. Currently, the field is rapidly switching from conventional algorithms such as Scale-Invariant Feature

Transform (SIFT) and Speeded-Up Robust Features (SURF) to deep learning and augmented reality-based computer vision.

3.2.3 Ontology

The foundational notion of the term "Ontology" in the Computer Science and AI research field obeyed a standard definition, "an ontology is a formal, explicit specification of a shared conceptualization", where each facet has its unique attributes [43]. Until now, Ontologies have been categorised into four significant types depending on the intents and distinct granularity levels: Top-level/foundational ontology, Domain ontology, Task ontology and Application ontology [44]. As a whole, ontology research empowers the knowledge/ontology engineering branch by allowing them to discover how concepts exist, are linked together, and are used to reduce overall system costs by enhancing efficiency or quality. Ontology with ML contributes to software engineering, data analysis (capturing complexities between entities and relationships using automated information extraction), making recommender systems/portals or ontology-extended browsers, and providing computational support for data.

3.2.4 Knowledge graph (KG)

The idea of a knowledge graph (KG) getting lots of recognition from the global AI research community as well as the leading enterprises; as directed to Bagchi [45], "Knowledge graphs is the go-to solution for populating, reasoning and visualising knowledge domains in recent semantic information systems", or more specifically, "a graph of data intended to accumulate and convey knowledge of the real world, whose nodes represent entities of interest and whose edges represent relations between these entities" [46]. KG has become a specialised interdisciplinary domain of study that uses graph theory, mathematical logic and reasoning, human-machine interaction, knowledge and data representation, and cognitive and semantic modelling, and constantly evolves using ML and DL algorithms trends to archive beyond representation. As the field is still under development, the definitions are contentious, and from the perspective of construction, two types of KGs have been explored by the researchers till now, (i) *"data level KGs"* or *"entity graphs (EGs)"* and (ii) *"schema level KGs"* or *"entity type graphs (ETGs)"* [47].

4. Open Science with AI: the promising future

There are different schools of thought (Public School, Democratic School, Pragmatic School, Infrastructure School, and Measurement School) have seen and defined open science from different perspectives [48], its core foundation embedded in the formulation of such robust standards and sustainable policy frameworks in order to accomplish the desired infrastructure for academic freedom, collaboration and practices and overcoming the challenges of legal, technological, cultural, and ideological transformation. In the 41st General Conference of United Nations Educational, Scientific and Cultural Organisation [6], the advisory committee issued a draft recommendation on Open Science comprising a consistent definition of open science, four core values and six guiding principles, and action guidance under seven key areas. Based on the "online information meeting on Implementation of the UNESCO Recommendation on Open Science" [49], ad hoc working groups highlighted the role of

Open Science in scientific prosperity, especially as a critical accelerator for the implementation of all of the Sustainable Development Goals (SDGs) [50], as well as bridging the science, technology, and innovation gaps between and within countries, and pointed out four key pillars such as Open scientific knowledge, Open Science infrastructure, Open engagement of societal actors, and Open dialogue with other knowledge systems and addressed the challenge of achieving them. They are also developing a series of supporting tools (Open Science Toolkit+Open Science Info App) [51, 52], technical briefs, fact sheets, and guidelines that are easy to use, reuse, expand, and update and accessible to all. AI plays a game-changing role in attaining all four key pillars and strategic implementation of the recommendations, including monitoring/analysing the progress and, most importantly, making the technical dreams a reality.

European Commission [5] monitors and equips trends, data and indicators related to global Open Science advances and functions with two expert groups (Open Science Policy Platform and expert group on indicators) now became an integrated part of the "Horizon Europe Programme of 2021 for research and innovation continues developing the Open Science policy framework under the section dubbed "8 ambitions of the EU's open science policy", also highlighted by the League of European Research Universities' "eight pillars of Open Science" [53] as correctly identified by Bagchi [54]. All eight ambitions and UN SDGs are a component of Horizon Europe, including Research Infrastructure with the European Open Science Cloud (EOSC), Marie Sklodowska-Curie Actions, and the Open Research Europe (ORE) publishing platform, which all have AI implications. Regarding this, OpenAIRE (https://www.openaire.eu/), a socio-technical infrastructure/legal entity, supports EC and European Open Science mandates by taking care of its policy alignment through the National Open Access Desks network (NOADs) and building several cutting-edge services through Open Science Graph in the backend, which connects worldwide infrastructures and networks to disseminate open research findings [55]. The Novel EOSC Services for Emerging Atmosphere, Underwater and Space Challenges [56], an ambitious initiative with AI services for Open Science, was introduced to develop an additional application/resolving the existing issue(s) within EOSC. IntelComp's (https://intelcomp.eu/) revolutionary Cloud Platform delivers AI-based services or "*Policy Intelligence*" to EU's public administrators and policymakers for data and evidence-driven policy creation in Science, Technology, and Innovation (STI) policy. Under the leadership of partner "Spanish Secretary of State for Digitalization and Artificial Intelligence" (SEDIA), the Project examines the co-development and Platform's usage for AI R&D intelligence gathering and the industry's AI credentials.

Center for Open Science (COS), aimed at reducing the waste and discovering knowledge solutions and cures for the world's most pressing needs, offers the Open Science Framework (OSF) [57], an open-source and freely accessible project management tool for open, reproducible, and trustworthy research practices in all stages of the research life cycle. The OSF infrastructure supports cultural shifts by enabling rigour and transparency across the research life cycle. It has mastered the art of reproducibility and preregistration, which is the process of developing research questions and an analysis strategy before viewing the study's findings [58]. Besides these, it also has cloud-based archival solutions, preprint vaults, integration of local repositories, collection management through customisable filtering, and taxonomies with robust discovery and retrieval. The validity of AI discoveries depends on reproducible experiments and Open Science with FAIR principles [59] of distributing data, software, and other scientific resources in public repositories under liberal licences, which would be advantageous for the AI research community.

The Organisation for Economic Co-operation and Development [60] revised its policy framework in order to incorporate new technologies and guiding principles reflected in all seven key areas, namely Data governance for trust, Technical standard and practices, Incentives and rewards, Responsibility, ownership and stewardship, Sustainable infrastructure, Human capital and International cooperation for access to research data. According to the January 2021 mandate, the revised template entitled "Recommendation of the Council concerning Access to Research Data from Public Funding" in which the expanded scope covers research data, metadata, bespoke algorithms, workflows, models, software and code.

5. Discussion

The synergy between AI and open science has the potential to transform the way scientific research is conducted, disseminated, and applied since it draws from both the cognitive learning processes of individuals and the learning processes of enterprises during open innovation.

Open science and AI can work together:

a. *Data analysis and interpretation*: AI can assist in processing and analysing large datasets quickly and efficiently. Researchers can use AI algorithms to identify patterns, trends, and correlations within vast amounts of data, leading to faster insights and discoveries.

b. *Automated experimentation*: AI-powered bots and systems can conduct experiments autonomously, increasing the efficiency of data collection and minimising human errors. Researchers can focus more on designing experiments and interpreting results.

c. *Text and literature analysis*: AI-driven NLP can sift through vast amounts of scientific literature, extracting relevant information, summarising articles, and identifying gaps in research. This enhances researchers' ability to stay updated and build on existing knowledge.

d. *Collaboration and crowd science*: Open science promotes collaboration among researchers globally. AI-powered platforms can facilitate sharing of data, methodologies, and findings, enabling researchers to collaborate on a larger scale.

e. *Reproducibility and transparency*: AI can help ensure the reproducibility of research results by providing detailed documentation of analysis methods and codes used in studies. This enhances transparency and trust in research outcomes.

f. *Data sharing and accessibility*: AI can aid in organising and structuring research data, making it more accessible and reusable. This contributes to the overall integrity and validity of scientific findings.

g. *Education and outreach*: AI can facilitate the creation of interactive educational materials, simulations, and virtual labs, making scientific concepts more engaging and accessible to learners of all ages.

AI-powered tools have the potential to both positively and negatively impact the quality of science. While AI can enhance various aspects of scientific research and discovery, it can also introduce challenges and concerns that need to be carefully managed. For instance, the attributes mentioned above can positively affect the quality of science in terms of efficiency, speed, data analysis, pattern recognition, reproducibility, and hypothesis generation. But it is necessary to take into account the potential negative impacts and concerns:

a. *Bias and fairness*: Biases in the training data might be inherited by AI algorithms and result in unreliable predictions. If not properly addressed, these biases can affect the quality and fairness of research results.

b. *Data overfitting*: AI models, if not carefully designed and validated, can overfit the training data, leading to results that do not generalise well to new data. This can undermine the validity of research findings.

c. *Lack of interpretability*: Deep learning models, in particular, can be challenging to interpret. This can make it difficult to understand the reasoning behind AI-generated results, potentially affecting the trustworthiness of those results.

d. *Data privacy and security*: AI often requires access to large, sensitive datasets. Ensuring the privacy and security of this data while still conducting meaningful research can be challenging.

e. Dependency on technology: Overreliance on AI-powered tools might reduce the emphasis on critical thinking and traditional scientific methodologies, potentially undermining research rigour. For example, AI-powered text generators (OpenAI's GPT-3, Google's ChatGPT, Microsoft's XiaoIce), image generators (OpenAI's DALL-E, Google's DeepDream, NVIDIA's StyleGAN, StyleGAN2, BigGAN) became so prevalent that enterprises have begun to prioritise these over the human brain.

f. *Loss of serendipity*: AI may prioritise known patterns and trends, potentially reducing the likelihood of serendipitous discoveries that arise from creative thinking and unexpected observations.

g. *Ethical considerations*: The use of AI in research raises ethical concerns related to consent, transparency, and the potential for unintended consequences. Such issues may occur in the form of digital divides (exacerbate existing disparities in access to information, resources, and opportunities), invisibilisation of certain groups (reinforce dominant norms and exclude marginalised voices and perspectives), and discrimination of minorities (bias present in training data) because of the nature of algorithmic transparency and accountability. Therefore, ensuring ethical research practices is crucial.

It's vital to remember that the degree to which AI improves scientific quality is not solely determined by the technology itself but also by how researchers, institutions, and the broader scientific community use and integrate AI into their workflows. It is essential to strike a balance between automation and human oversight, adopt rigorous validation practises, promote transparency, and continuously address ethical and bias-related challenges, ethical considerations, data privacy, bias mitigation, and standardisation of practises in order to fully leverage the synergy between AI and open science and mitigate potential negative impacts and harness the positive potential of AI. Nevertheless, the combined power of AI and open science has the potential to accelerate the pace of discovery, democratise access to knowledge, and foster a more collaborative and innovative scientific community.

6. Conclusions

Open Science, as the pristine engine of prosperity (viewed as better science), developed and improved through continuous research (technological boon making it a reality) and levelling up towards the hope of wisdom of which the homo sapience culture must acquire for its survival. The last two decades have seen a staggering increase in data, which, in tandem with the world's changing technology, particularly in the form of AI, puts the greatest challenge of our time to one side and, on the other hand, opens up new potential to build a continuous knowledge base. Throughout the study, we have shown numerous use case scenarios of how Open Science symbioses with AI accelerate the discovery of knowledge, collaboration, and solutions to achieve them, as well as address the potential threats and ethical concerns that require a multi-faceted approach involving cooperation among governments, tech companies, researchers, ethicists, and affected communities to ensure that AI technologies are developed and deployed in ways that respect human rights, promote fairness, and avoid exacerbating existing inequalities. One thing that needs special mention here is that AI plays a significant role in executing Open Science practices, and two distinct notions with multifarious attributes amalgamating, future work on recommending holistic policy guidelines must be carried out to detour the ethical issues related to them.

Acknowledgements

It is the result of a personal research endeavour that led to this work.

Conflict of interest

The authors declare no conflict of interest.

Author details

Mayukh Sarkar[1]* and Sruti Biswas[2]

1 Saint Claret College, Ziro, India

2 IQ City United World School of Business, Kolkata, India

*Address all correspondence to: mayukhs@drtc.isibang.ac.in

References

[1] Segan C. Billions and Billions: Thoughts on Life and Death at the Brink of the Millennium. New York: Ballantine Books; 1997

[2] Open Knowledge Foundation. The Open Definition [Internet]. 2022. Available from: https://opendefinition.org/

[3] Hanwell MD. What is Open Science [Internet]. 2022. Available from: https://opensource.com/resources/open-science#:~:text=Open%20science%20arguably%20began%20in,such%20as%20the%20Royal%20Society

[4] FOSTER. Open Science Definition [Internet]. 2022. Available from: https://www.fosteropenscience.eu/foster-taxonomy/open-science-definition

[5] European Commission. Open Science [Internet]. 2022. Available from: https://ec.europa.eu/info/sites/default/files/research_and_innovation/knowledge_publications_tools_and_data/documents/ec_rtd_factsheet-open-science_2019.pdf

[6] UNESCO. Draft Recommendation on Open Science [Internet]. 2021. Available from: https://unesdoc.unesco.org/ark:/48223/pf0000378841

[7] Britannica. Human Intelligence [Internet]. 2022. Available from: https://www.britannica.com/science/human-intelligence-psychology

[8] Poole DL, Mackworth AK. Artificial Intelligence: Foundations of Computational Agents. 2nd ed. Cambridge: Cambridge University Press; 2017

[9] Swartz A. Guerilla Open Access Manifesto [Internet]. 2008. Available from: https://archive.org/download/GuerillaOpenAccessManifesto/Goamjuly2008.pdf

[10] Bodó B. The genesis of library genesis: The birth of a global scholarly shadow library. In: Karaganis J, editor. Shadow Libraries: Access to Knowledge in Global Higher Education. Cambridge: The MIT Press; 2018. pp. 25-51

[11] Sci-Hub. Elbakyan [Internet]. 2023. Available from: https://sci-hub.se/alexandra#works

[12] Elbakyan A, Bohannon J. Data from: Who's downloading pirated papers? Everyone [dataset]. Dryad. 2021. DOI: 10.5061/dryad.q447c

[13] Bodó B. Library genesis in numbers: Mapping the underground flow of knowledge. In: Karaganis J, editor. Shadow Libraries: Access to Knowledge in Global Higher Education. Cambridge: The MIT Press; 2018. pp. 53-77

[14] Houle L. Sci-Hub and LibGen: What if... why not? In: IFLA World Library and Information Congress 2017 – Wrocław, Poland – Libraries. Solidarity. Society. Gdansk. 2020. Available from: http://library.ifla.org/id/eprint/1892/1/S12-2017-houle-en.pdf

[15] Valladares-Garrido MJ et al. Association between the use of Sci-Hub and consultation of scientific journals by medical students from six Latin American countries: A secondary analysis. Heliyon. 2023;**9**(e17868):1-11. DOI: 10.1016/j.heliyon.2023.e17868

[16] Ajani YA, Tella A, Okere S. Access to full-text documents in libraries via Sci-Hub: A blessing in disguise to library users. Library Hi Tech News. 2023:1-4. DOI: 10.1108/LHTN-03-2023-0053 [Ahead-of-print]

[17] Sci-Hub. Stats [Internet]. 2023. Available from: https://sci-hub.se/stats

[18] Palmer M. Data Is the New Oil [Internet]. 2006. Available from: https://ana.blogs.com/maestros/2006/11/data_is_the_new.html

[19] Tang J, Zhang J, Yao L, Li J, Zhang L, Su Z. ArnetMiner: Extraction and mining of academic social networks. In: Proceedings of the 14th ACM SIGKDD International Conference on Knowledge Discovery and Data. New York: Association for Computing Machinery; 2008. pp. 990-998. DOI: 10.1145/1401890.1402008

[20] Ecer D, Maciocci G. ScienceBeam - Using Computer Vision to Extract PDF Data [Internet]. 2017. Available from: https://elifesciences.org/labs/5b56aff6/sciencebeam-using-computer-vision-to-extract-pdf-data

[21] Github. Elifeciences/sciencebeam-parser [Internet]. Available from: https://github.com/elifesciences/sciencebeam-parser

[22] Github. Elifeciences/peerscout [Internet]. Available from: https://github.com/elifesciences/peerscout/

[23] Ecer D, Shannon P. AI for automation and influence in open science publishing. In: Implementing AI. London: Artificial Intelligence Conference; 2018. Available from: https://conferences.oreilly.com/artificial-intelligence/ai-eu-2018/public/schedule/detail/70119.html

[24] Wang K. Opportunities in open science with AI. Frontiers in Big Data. 2019;**2**(26):1-4. DOI: 10.3389/fdata.2019.00026

[25] Sinha A, Shen Z, Song Y, Ma H, Eide D, Hsu B, Wang K. An overview of Microsoft academic service (MAS) and applications. In: Proceedings of the 24th International Conference on World Wide Web. New York: Association for Computing Machinery; 2015. pp. 243-246. DOI: 10.1145/2740908.2742839

[26] Wang K, Shen Z, Huang C, Wu C, Eide D, Dong Y, et al. A review of Microsoft academic services for science of science studies. Frontiers in Big Data. 2019; **2**(45):1-16. DOI: 10.3389/fdata.2019.00045

[27] Zhang F, Liu X, Tang J, Dong Y, Yao P, Zhang J, et al. OAG: Toward linking large-scale heterogeneous entity graphs. In: Proceedings of the Twenty-Fifth ACM SIGKDD International Conference on Knowledge Discovery and Data Mining. New York: Association for Computing Machinery; 2019. pp. 2585-2595. DOI: 10.1145/3292500.3330785

[28] McCarthy J. What Is a Artificial Intelligence [Internet]. 2007. Available from: http://jmc.stanford.edu/articles/whatisai/whatisai.pdf

[29] Kubat M. An Introduction to Machine Learning. 3rd ed. Cham: Springer; 2021

[30] El Naqa I, Murphy MJ. What is machine learning? In: El Naqa I, Li R, Murphy MJ, editors. Machine Learning in Radiation Oncology: Theory and Applications. Cham: Springer; 2015

[31] Wani A, Khoshgoftaar TM, Palade V. Deep Learning Applications. Vol. 2. Singapore: Springer; 2021

[32] Zhang C, Lu Y. Study on artificial intelligence: The state of the art and future prospects. Journal of Industrial Information Integration. 2021;**23**: 100224. DOI: 10.1016/j.jii.2021.100224

[33] Sutton RS, Barto AG. Reinforcement Learning: An Introduction. 2nd ed. Cambridge: MIT Press; 2018

[34] Szepesvári C. Algorithms for Reinforcement Learning [Internet]. 2019. Available from: https://sites.ualberta.ca/~szepesva/papers/RLAlgsInMDPs.pdf

[35] Gridin I. Learning Genetic Algorithms with Python. New Delhi: BPB Publications; 2021

[36] Kramer O. Genetic Algorithm Essentials. Cham: Springer; 2017

[37] Kotyrba M, Volna E, Habiballa H, Czyz J. The influence of genetic algorithms on learning possibilities of artificial neural networks. Computers. 2022;**11**(5):70. DOI: 10.3390/computers11050070

[38] Tao J, Zhang R, Zhu Y. DNA Computing Based Genetic Algorithm: Applications in Industrial Process Modeling and Control. Singapore: Springer; 2020

[39] Yu X, Gen M. Introduction to Evolutionary Algorithms. London: Springer-Verlag; 2010

[40] Patel AA, Arasanipalai AU. Applied Natural Language Processing in the Enterprise: Teaching Machines to Read, Write, and Understand. Beijing: O'Reilly; 2021

[41] Tunstall L, von Werra L, Wolf T. Natural Language Processing with Transformers: Building Language Applications with Hugging Face. Beijing: O'Reilly; 2022

[42] Bekhit AF. Computer Vision and Augmented Reality in iOS: OpenCV and ARKit Applications. New York: Apress; 2022

[43] Studer R, Benjamins R, Fensel D. Knowledge engineering: Principles and methods. Data & Knowledge Engineering. 1998;**25**(1–2):161-197. DOI: 10.1016/S0169-023X(97)00056-6

[44] Staab S, Studer R, editors. Handbook on Ontologies. 2nd ed. Dordrecht: Springer; 2008

[45] Bagchi M. A large-scale, knowledge-intensive domain-development methodology. Knowledge Organization. 2021;**48**(1):8-23. DOI: 10.5771/0943-7444-2021-1-8

[46] Hogan A et al. Knowledge Graphs [arXiv]. 2021. Available from: https://arxiv.org/pdf/2003.02320.pdf

[47] Giunchiglia F, Bocca S, Fumagalli M, Bagchi M, Zamboni A. iTelos - Building Reusable Knowledge Graphs [arXiv]. 2021. Available from: https://arxiv.org/pdf/2105.09418.pdf

[48] Fecher B, Friesike S. Open science: One term, five schools of thought. In: RatSWD Working Paper. Vol. 218. 2014. Available from: https://www.econstor.eu/bitstream/10419/75332/1/746340028.pdf

[49] UNESCO. Online Information Meeting on Implementation of the UNESCO Recommendation on Open Science [Internet]. 2022. Available from: https://www.youtube.com/watch?v=Yw9U4mwGVTE

[50] United Nations. Sustainable Development Goals [Internet]. Available from: https://sdgs.un.org/goals

[51] UNESCO. Open Science Toolkit [Internet]. Available from: https://www.unesco.org/en/open-science/toolkit

[52] UNESCO. UNESCO index of Open Science Knowledge Sharing Platforms [Internet]. Available from: https://www.unesco.org/en/open-science/knowledge-sharing

[53] LERU. Open science and its role in universities: A roadmap for cultural change. Advice Paper, 24. 2018. pp. 1-32. Available from: https://www.leru.org/files/LERU-AP24-Open-Science-full-paper.pdf

[54] Bagchi M. Open science for an open future. In: Madalli DP, Prasad ARD, editors. Proceedings of the International Conference on Exploring the Horizons of Library and Information Sciences: From Libraries to Knowledge Hub. Bangalore: Documentation Research and Training Centre, Indian Statistical Institute; 2018. pp. 422-431

[55] Manola N, Rettberg N, Manghi P, Mertens M, Schmidt B, Steiner T, et al. Achieving Open Science in the European Open Science cloud: Setting out OpenAIRE's vision and contribution to EOSC. OpenAIRE MAKE. 2019. DOI: 10.5281/zenodo.3610132

[56] NEANIAS. AI Services for Open Science [Internet]. 2021. Available from: https://www.neanias.eu/images/neanias/Articles/202102_WP4_AI_Services_for_Open_Science.pdf

[57] Center for Open Science. OSF [Internet]. Available from: https://osf.io/4znzp/

[58] Nosek BA, Ebersole CR, DeHaven AC, Mellor DT. The preregistration revolution. PNAS. 2017;**115**(11):2600-2606. DOI: 10.1073/pnas.1708274114

[59] GO FAIR. FAIR Principles [Internet]. Available from: https://www.go-fair.org/fair-principles/

[60] OECD. Recommendation of the council concerning access to research data from public funding, OECD/LEGAL/0347. 2022. Available from: https://legalinstruments.oecd.org/api/print?ids=159&lang=en